Bell Telephone System's Preeminent Role in the Growth of Industrial Design

Bell Telephone System's Preeminent Role in the Growth of Industrial Design

Ralph O. Meyer
Russell A. Flinchum

Purdue University Press • West Lafayette, Indiana

Cataloging-in-Publication Data is on file at the Library of Congress.
978-1-62671-108-2 (hardback)
978-1-62671-109-9 (paperback)
978-1-62671-110-5 (epub)
978-1-62671-111-2 (epdf)

Images throughout this publication provided by author Ralph O. Meyer unless otherwise indicated.

COVER IMAGES

Top row, left to right:

Bell Labs reproduction of first commercial handheld telephone, 1877. Photo by Ted Mueller.

Western Electric No. 202 handset desk stand, 1930. Photo by Matthew Gay. Courtesy of
    Gregg Museum of Art & Design, North Carolina State University.

Western Electric No. 302 complete telephone, 1937. Photo by Sally Andersen-Bruce.
    Courtesy of the US Postal Service.

Middle row, left to right:

Western Electric No. 500 post-WWII telephone, 1949. Photo by Matthew Gay. Courtesy of
    Gregg Museum of Art & Design, North Carolina State University.

Outline of Western Electric E-type handset with dimensions and angles.

Western Electric Princess bedroom extension telephone, 1959. Photo by Richard Rose.

Bottom row, left to right:

Western Electric No. 2500 Touch-Tone telephone, 1963. Photo by Matthew Gay. Courtesy of
    Gregg Museum of Art & Design, North Carolina State University.

Western Electric Trimline dial-in-handset telephone, 1965. Photo by Matthew Gay.
    Courtesy of Gregg Museum of Art & Design, North Carolina State University.

AT&T No. 700 business telephone set, 1984. Photo by Paul Fassbender.

In cheerful memory of Sue, author Meyer's wife of 63 years. She is seen here at the operator's board of the Chapel Hill Telephone Company in 1961, at which time one could dial "o" and reach a local operator to get someone's phone number and street address or to get assistance in an emergency.

CONTENTS

ACKNOWLEDGMENTS

We are indebted to Donald Genaro, one of Henry Dreyfuss's principal designers, for the extensive information he provided earlier—some of which has made its way into this book. Paul Fassbender, a prominent American telephone historian, was immensely helpful. Remco Enthoven, curator of the American Museum of Telephony, made measurements, took scaled photographs, and provided some document scans. It's impossible to overstate the research value of several hundred historically inclined members of Telephone Collectors International (USA), Telecommunications Heritage Group (Britain), and Sammler- und Interessengemeinschaft für das historische Fernmeldewesen (Germany). Many of their members provided assistance—especially the following: Dag Ståle Karlsen gave us a 1931 Ericsson telephone and a lot of information about this Norwegian design. Jeffrey Lamb donated a Northern Electric Uniphone and provided much information about Canadian telephones. Bob Freshwater helped substantially with information about British telephones. Dietrich Arbenz was very helpful regarding Siemens & Halske designs, and his work is cited in the text. Karl Brose, a chemist, provided significant assistance including information about plastics. Jack Ryan provided general information on international designs. Ole Warrelmann, a newsletter editor, and Gerhard Fuchs, an electrical engineer, provided information and many documents on European telephone designs. Gary Goff, Steve Hilsz, and Richard Rose provided specimens for examination. John Infurna, a machinist, helped us understand how Bell's box telephone was manufactured. Thomas Adams provided access to his collection. Others contributed photographs and are identified in the figure captions. We are genuinely grateful for the generous help we received. Finally, we want to thank author Meyer's son and daughter—and the publisher's reviewers—for critical comments on various drafts that resulted in revisions and made the book much more readable than our original manuscript.

INTRODUCTION

H istories of design pay scant attention to the corded telephone, which played an immeasurable role in earlier communications. When it is addressed, credit is often given to well-known designers for the work of others, and the discipline of industrial design within the Bell Telephone organization is unnoticed.[1] Here, we try to set the record straight. This book digs deep into industrial design of the telephone from its beginning to the end of prominence of these older instruments.

Although the Bell System, with many Nobel Prizes, is justly acknowledged for its technical prowess, it should also be known for its early and considerable impact on the developing discipline of industrial design. Unrecognized and not yet named, industrial design appeared in the first generation of commercial telephones. A radical change in the shape of the telephone's receiver was made by William Channing and his friends at Brown University. This attractive—and

---

1 The original Bell Telephone Company developed over time into the Bell Telephone System, or simply the Bell System. At its peak, it comprised the American Telephone & Telegraph Company (AT&T) for corporate management and long-distance lines, Bell Telephone Laboratories (Bell Labs) for research and development, the Western Electric Company (Western Electric) for manufacturing, and more than 20 local operating companies for telephone service (e.g., Cincinnati and Suburban Bell Telephone Company).

now familiar—shape was adopted by the original Bell Telephone Company. The emerging Bell System later embraced the arts under the guidance of its revered president Theodore Vail. After a period without further attention to aesthetics, however, an effort to design a new telephone failed when outside artists were called in, and Harold Dodge, a Bell Labs staff member with artistic interests, was able to provide the needed appearance improvements. A tug of war ensued between executives at AT&T and the management at Bell Labs and led to a successful formula by which designers were embedded with engineers. At this point young Henry Dreyfuss, who would later become known as a pioneer of industrial design, was retained as a design consultant, and Dreyfuss matured under the influence of Bell Labs. With his help, the pre–World War II No. 302 telephone was designed by George Lum at Bell Labs, and the phone became iconic. The postwar No. 500 telephone, designed together by Henry Dreyfuss and Robert Hose from Bell Labs, was so successful its design was copied around the world. And the final major design for the Bell System was the Trimline telephone, sculpted by Donald Genaro of the Dreyfuss firm—a telephone that today is in the collection of the Museum of Modern Art. With substantial input from Bell Labs engineers, the telephone designs produced by Dreyfuss and his staff were ultimately manufactured and installed in more than 100 million American businesses and homes.

PART 1

INDUSTRIAL DESIGN
TAKES ROOT

B y the year the first practical telephone went into service, Colorado had just become the 38th state and the United States spread from Maine to California. Coast-to-coast telegraph service was in place, but sending messages was painfully slow—letter by letter—and required trained operators who could understand the click, click, clicks of Morse code. The Civil War had been over for little more than a decade. Manufacturing capabilities in the North had survived and the industrial revolution had flourished. Technical elements were in place, and the telephone's time had come.

Maxwell's unified equations for electricity and magnetism were recent (1861–1862) and embodied Ampere's law and Faraday's law. Ampere's law describes a magnetic field that is produced when an electric current flows through a wire. Faraday's law describes a current in a wire that is caused by a changing magnetic field. The latter had been independently discovered by Michael Faraday in Britain and Joseph Henry in America, and Bell had met with Henry for advice. Bell understood Faraday's law and used its principle for his telephone. Except for physicists of the day, few understood how Bell's telephone worked, yet that would be necessary to apply true industrial design.

1

ALEXANDER GRAHAM BELL'S TELEPHONE IN A BOX

The world's first practical telephone, Alexander Graham Bell's box phone, went into service in April 1877. It used the same instrument for a transmitter and a receiver, had a range of more than 100 miles, and needed no batteries. The telephone was enclosed in a box (fig. 1.1) that was big at 9 inches by 12 inches (23 cm by 30 cm) and heavy at almost 12 pounds (5.4 kg)—and it was awkward to use because one had to talk into the opening in the box and then turn around and listen at the same opening. (Incidentally, scientists at Bell Labs used metric units whereas engineers in the same organization used inches and pounds because that's how American machine tools were calibrated during this period of history.)

The operation of Bell's box telephone was sophisticated, technically beautiful, and rarely understood, utilizing two important principles of physics (Meyer 2020; 2022, 11). Bell's concept was for pressure waves from spoken words to vibrate a magnetized diaphragm in front of telegraphers' coils and thereby generate an electric current according to Faraday's law. The electric current would be undulating, in Bell's words, increasing and decreasing with the frequency and amplitude of the diaphragm's vibrations. The telephone would work in reverse as a receiver according to Ampere's law.

FIGURE 1.1 The world's first commercial telephone, Bell's 1877 box telephone. (Photo courtesy of the AT&T Archives and History Center.)

Figure 1.2 shows a functional diagram of Bell's concept. A soft iron diaphragm became the magnet that vibrated in front of coils. Magnetism of the diaphragm and soft iron pole pieces, which were used as coil cores, was induced by a strong hardened steel permanent magnet. Telegraph coils, already in existence, were generally used in pairs, so it is not surprising that Bell used a pair of coils in his design. The instrument was also referred to as a magneto telephone because it generated its own electricity.

The individual components in the actual box telephone were themselves large and seemed irreducible in size (fig. 1.3). Any change in external shape that might

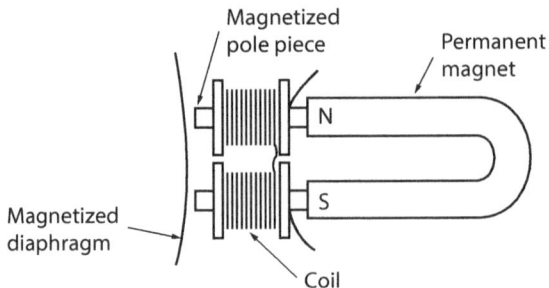

FIGURE 1.2 Diagram (top view) of Bell's 1877 box telephone.

FIGURE 1.3 Interior of Bell Labs replica of Bell's 1877 box telephone. (Photo by Ted Mueller.)

make the telephone more convenient would require an understanding of its operating principles, and any redesign would have to be done from the inside out. The telephone shown in figure 1.3 is actually a replica that was made in the famed Model Shop on West Street in New York (Bell Labs 1925). By the way, the permanent magnet is seen to be made of four thin horseshoe magnets that were easier to magnetize at that time, but the layers had no special properties.

2

AN UNEXPECTED
CONFIGURATION CHANGE

lexander Graham Bell was working in Boston, but down in Providence, Rhode Island, experiments were being conducted at Brown University on Bell's telephone concept. The research was being done out of pure scientific interest, along with the hope of contributing something to its perfection, and the work was headed by William Channing (Channing 1877).

William Francis Channing (1820–1901) was an interesting character by any measure. Born in Boston, he was the son of Rev. William Ellery Channing (a founder of Unitarianism) and Ruth (Gibbs) Channing (Bruce 1990, 225; Johnson and Brown 1904; Kelly and Burrage 1920). During the abolition movement, he was a leader among activists. He began higher education at Harvard, but then decided to study medicine at the University of Pennsylvania in Philadelphia, where he received an MD. Channing later teamed up with Moses Farmer in Boston to develop a fire alarm telegraph system (Channing and Farmer 1857). He subsequently moved to Providence and hung around the Physics Department at Brown University, where they were reproducing Bell's work.

The group in Providence would have been aware of Bell's concept from his 1876 patent, and they would have known the details of his telephone from his patent of January 1877 (Bell 1876b, 1877). These experimenters were using one pole of a bar magnet rather than both poles of a horseshoe magnet as Bell was using. The operating principle was the same, however, and the monopole device also

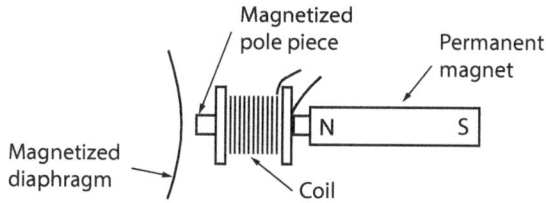

FIGURE 2.1 Diagram (side view) of Channing's 1877 handle telephone.

worked without batteries. Figure 2.1 shows a functional diagram of Channing's telephone, which can be compared with figure 1.2 to see the equivalence. In the end Bell's box phone, with its stronger bipolar magnet and larger diameter diaphragm, produced stronger electrical signals than Channing's telephone, but this monopole device had qualities of appearance and convenience that would be profoundly useful.

Figure 2.2 shows a drawing of a cross section of Channing's handle telephone that was constructed in early May 1877 (Prescott 1884, 102–4). Components in this drawing can be compared with the schematic diagram in figure 2.1. In his July 16 letter to Bell, Channing said three such handle telephones were given in May to Bell's assistant, Thomas Watson, to pass along to Bell later that month (Channing 1877). Bell had been independently experimenting with a U-shaped magnet in a bulky handle, but when he received a pair of handle telephones from Channing, he switched to using a straight magnet.

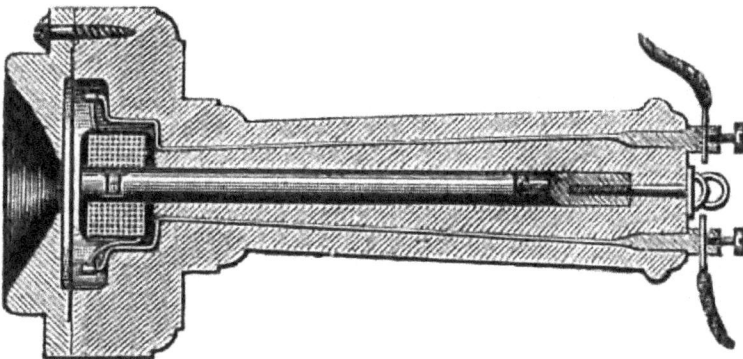

FIGURE 2.2 Cross-sectional drawing of Channing's handle telephone, early May 1877. (Reprinted from Prescott 1884, 104.)

Channing's group of experimenters consisted of Eli W. Blake, professor of physics at Brown; John Peirce, professor emeritus at Brown; Edison P. Jones, who prepared drawings; Louis W. Clarke and Charles E. Sustin, who took part in the construction and testing of instruments; and Channing himself (Bell 1908, 431). The Providence experimenters had made thousands of observations and experiments during the previous six months in the physics laboratory at Brown before arriving at their final design. Channing's connection to Bell was via Farmer, who had the fire alarm telegraph system work done in Charles Williams's shop in Boston. Farmer met Bell through Watson, who worked there, and Farmer had become an informal advisor to Bell.

Channing, in his lengthy letter to Bell in July 1877, made statements that clearly show that industrial design—even without a name—was being applied. In the middle of Channing's letter, he says, "I have taken pleasure in the assurance that the miniature or 'baby' telephone & mouthpiece of April, and the 'handle telephone' of May, have been material helps in making the telephone portable, practical, & commercially successful" (Channing 1877). Later in the letter Channing says, "What inventive merit it has is practical, putting the telephone in compact, elegant form for popular use, & economical, making the telephone easily usable and salable." These statements are almost the definition of industrial design.

Here, in the handle or hand telephone, is a perfect example of an approach that says, "Appearance should be developed from the inside out, not merely created as a mold into which the engineers would eventually squeeze the mechanism," as later stated by Henry Dreyfuss (Dreyfuss 1955, 102). Bell had tried to stuff his bipolar design into a handle, but it wasn't satisfactory. Channing and his physicist friends had to fully understand how Bell's telephone worked in order to redesign the inside before they could fit it into a convenient shell.

The work in Providence had been going on in the spring of 1877, during which time Bell and Watson were engaged in a series of lectures to promote the telephone. By summer, however, Bell shifted his attention from promotion to marriage, as he would get married in July and leave on an extended honeymoon in August. Bell married Mabel Hubbard, the daughter of his financial sponsor, Gardiner Hubbard (Bruce 1990, 234). Thus starting around May, credit for all design improvements must be given to Watson, whom Bell left in charge of technical matters. (And in July 1877, the Bell Telephone Company was established, so we will refer to the Bell Company—or later the Bell System—when not talking about Bell personally [Hubbard 1877]).

FIGURE 2.3 Bell Company hand telephone, June 1877.
(Photo courtesy of the Division of Work and Industry, National
Museum of American History, Smithsonian Institution.)

In June, Watson made several versions of a handle telephone, sometimes referred to as butter-stamp telephones, eventually settling on one that was very
similar to Channing's design. Watson's hand telephone, as it went into production (fig. 2.3), was 6 inches (15 cm) tall and weighed about a pound (0.5 kg). Wood
had been used in the construction of all of these early hand telephones, but wood
is not dimensionally stable with regard to age and climate change. Thus in mid-
December 1877, the Bell Company changed to hard rubber and made small shape
changes that would become classic.

Watson reported: "At first we used a single box telephone at a station, talking
and listening at the same orifice. But shifting from ear to mouth was bothersome
and we soon adopted the plan of installing two telephones at each station—a
box telephone fastened to the wall by the side of the call bell and a small portable telephone used as a receiver and held at the ear all the time one was speaking" (Watson 1926, 133). Watson went on to say that because the hand telephone
"didn't talk loudly enough to be used in the lectures, it was laid aside and [their]
principal attention given to the box form. When it became necessary to use two
telephones at a station, [he] revived the portable or hand form and put it into
shape for easy manufacture."

Keep in mind that the hand telephone and the box telephone are fundamen-
tally equivalent and either could be used for talking or listening. Figures 2.4–2.6
illustrate the various combinations that were considered. In figure 2.4, Bell is
speaking into a box telephone and Watson is listening to a box telephone. The

FIGURE 2.4 Bell and Watson using box telephones. (Illustration by Nicolas Waeckel.)

box telephone's bipolar magnet and large diameter diaphragm produced strong electrical signals when transmitting.

In figure 2.5, Bell is speaking into a hand telephone, which did not produce very strong electrical signals, and Watson is listening to a hand telephone. Box telephones are seen in the figure, but they are not being used.

FIGURE 2.5 Bell and Watson using hand telephones. (Illustration by Nicolas Waeckel.)

FIGURE 2.6 Bell and Watson speaking into box telephones and listening to hand telephones. (Illustration by Nicolas Waeckel.)

In figure 2.6, Bell is speaking into a box telephone while listening to a hand telephone. Watson is listening to a hand telephone and could speak into his box telephone at any time. This was the best combination, and subsequently the box telephone became known as a transmitter and the hand telephone as a receiver. This combination worked best because the transmitter needed to generate as much power as possible to get transmission over long distances, whereas the receiver would be held right against the ear such that relatively low sound levels could still be understood.

By the way, these drawings are examples of the artist-engineer combination that occurs often in industrial design. The artist, Nicolas Waeckel, studied sculpture as a child with Marcel Schoenenberger, a French sculptor, yet Waeckel's entire career was in nuclear power as a PhD mechanical engineer.

In August 1877, a wall-mounted version of the box telephone—then used only as a transmitter—was made by mounting the coils on a side of the permanent magnet near its tips rather than on the ends of the magnet (Prescott 1884, 441). This arrangement (fig. 2.7) is functionally the same as that in figure 1.2. After the first batch of box telephones was made, the layered permanent magnet was replaced by a solid horseshoe-shaped bar magnet. Instead of using clamps to hold the coils, such as shown in figure 1.2, the tips of the horseshoe magnet were softened by annealing then drilled and tapped, and ¼-inch (6.4 mm) pegs of soft iron were screwed into the tapped holes to serve as cores for the coils.

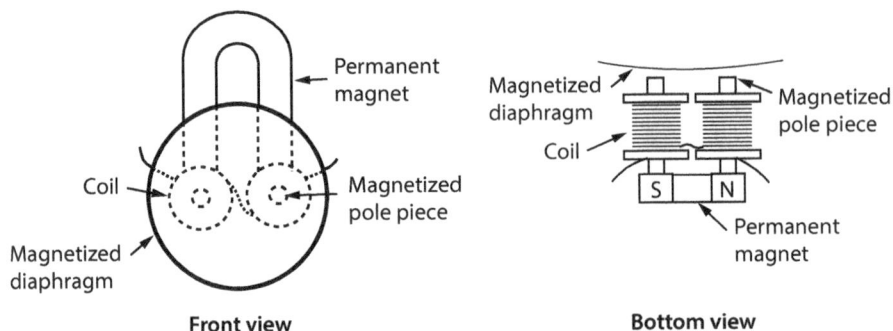

FIGURE 2.7 Diagrams of Bell's 1877 wall-mounted box telephone.

This more compact wall-mounted box phone (fig. 2.8), with its horseshoe magnet in the vertical plane, became the transmitting companion of the receiving hand telephone in installations by the new Bell Company.

FIGURE 2.8 Bell's 1877 wall-mounted box telephone used as a transmitter. (Photo courtesy of the AT&T Archives and History Center.)

3

RAPID DOMESTICATION

It is inevitable for every practical invention to have some rough edges, so to speak, when it is rushed into production. Early users would provide feedback, and designs would be modified quickly. This is a process we call design domestication, and it certainly took place with the original telephone.

If you think of the telephone as a talking version of the telegraph, which is how Alexander Graham Bell envisioned it, you might not anticipate the need to alert someone to an incoming call. When a telegraph sounder started clicking, everyone in the room could hear it and prepare to reply. Charles Williams, whose shop made the phones, had the first telephones installed between his house and his shop, but the new telephone was not loud enough to hail a listener. To simulate the clicks of a telegraph, Williams pecked at the diaphragm through the mouthpiece with a pencil to get attention at the other end (Watson 1940, 35). This of course damaged the soft iron diaphragm and brought the matter to Thomas Watson's attention.

Watson tried using a less damaging diaphragm thumper and then a buzzer, but neither was satisfactory. By autumn, Watson had developed a complete scheme involving a ringer, and on October 11, 1877, he filed a patent application for the arrangement shown in figure 3.1 (Watson 1878). Notice that Watson's scheme uses the same two line wires for ringing, talking, and listening—and a switch

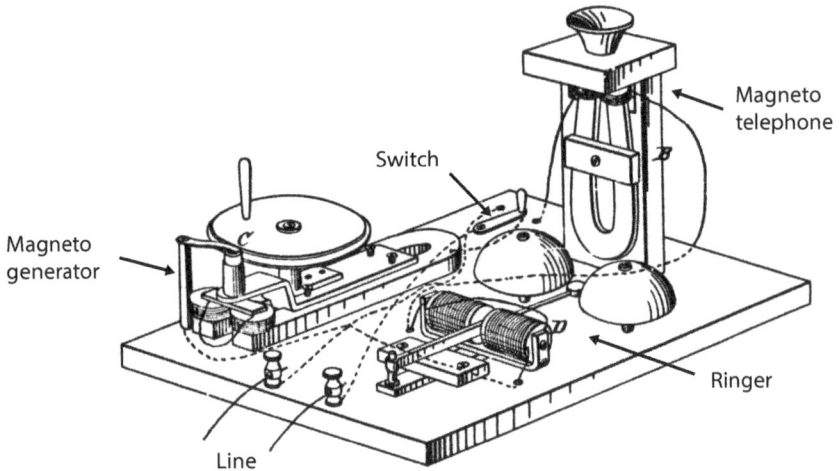

FIGURE 3.1 Diagram of the signaling arrangement from Watson's patent with labels added. (Adapted from Watson 1878.)

for changing the connection. This arrangement became the basis for all later telephone circuits, although the individual components changed significantly over time.

At Williams's shop, Watson's magneto generator for signaling, a ringer, and a switch were mounted on a box (fig. 3.2), along with electrical terminals for a transmitter, a receiver, and two line wires. A spark-gap lightning arrester was added as a safety measure, carried over from telegraph experience (Morey and Oehne 1908, 5). The arrester was connected between the lines, and a small gap was left such that the sawtooth plates did not touch and thus did not affect performance of the telephones. To protect the rest of the telephone from voltage surges, it was common practice to provide a way to short out the arrester in case a storm was expected. The small hole in the center of the arrester was tapered, and a tapered brass pin was provided to insert into that hole to make the electrical connection. (The pin was stored in a tapered hole at the very top of the box.) Because of the long and narrow shape of the box, these sets were referred to as "Williams' Coffins" (Watson 1940, 37).

In the fall of 1877, Hilborne Roosevelt filed an application for a patent on an automatic switch for telephone sets (Roosevelt 1879). It was a single-pole double-throw switch similar in function to the manual switch indicated above, but the movable contact was a spring that would normally touch the upper

FIGURE 3.2 Williams' Coffin set (*left to right:* complete, back of front panel, inside box), autumn 1877. (Photos by John Dresser.)

contact for talking. However, when the hand telephone was hung on the switch's hook, its weight would bend the spring down to touch the lower contact, thus switching the magneto and ringer across the line. The Bell Company acquired the rights to Roosevelt's switch and used it to replace the manual switch.

Figure 3.3 shows a later example of a Coffin phone with all of these improvements present. By this time, the box telephone was used only as a transmitter and the handheld telephone was used only as a receiver so they are labeled that way in the figure, where the receiver is seen to have the hard-rubber shell that was introduced in mid-December 1877 (Prescott 1884, 442). The lightning arrester's shorting pin is missing from this particular Coffin set, but the hole where the pin is usually stored can be faintly seen in figure 3.3.

The ringing mechanisms, hook switch, and lightning arrester could be viewed as engineering add-ons, but the handheld receiver was designed for user appeal and required a technical understanding to accomplish. The transmitter's configuration was also revised in the interest of the user, and this change, too, required technical knowledge. All of these features made the telephone safer, more comfortable to use, more efficient, and more likely to be sought by customers—the fundamental characteristics of industrial design.

Line terminals

Lightning arrester

Ringer gongs

Magneto crank

Switch

Transmitter

Receiver

FIGURE 3.3 Williams' Coffin set with labels added, ca. winter 1877–1878. (Photo courtesy of the AT&T Archives and History Center.)

In sharp contrast to this early period of domestication, the next 20 years of telephone development were almost all spent in pursuit of the goals of the telephone companies, without adding features for the customer. This 20-year period is covered in Part 2.

PART 2

A 20-YEAR DESIGN DROUGHT

For the next two decades, no changes were made in telephone design that we would call industrial design. Yes, there was some applied decoration, especially during the gay 1890s, but superficial decoration had been going on for ages in all sorts of venues, and it is not industrial design. For example, sewing machines and typewriters of the late 1800s often sported striping or decals. During this period of the telephone's evolution, however, all manufacturers were focused on performance and manufacturing with little attention to appearance or user friendliness. In the Bell Company, the effort was in pursuit of a "grand system."

4

THE GRAND SYSTEM
OF UNIVERSAL SERVICE

F our months to the day after the debut of the commercial telephone, Alexander Graham Bell and his bride left New York for a year-long honeymoon in Europe. He continued to promote commercialization of his telephones while he was there. For investors, he put together a business prospectus that envisioned a "grand system" (Bell 1878):

> In a similar manner, it is conceivable that cables of telephone wires could be laid underground, or suspended overhead, communicating by branch wires with private dwellings, country houses, shops, manufactories, etc., etc., uniting them through the main cable with a central office where the wires could be connected as desired, establishing direct communication between any two places in the city. Such a plan as this, though impracticable at the present moment, will, I firmly believe, be the outcome of the introduction of the telephone to the public. Not only so, but I believe, in the future, wires will unite the head offices of the Telephone Company in different cities, and a man in one part of the country may communicate by word of mouth with another in a distant place. I am aware that such ideas may appear to you Utopian and out of place, for we are met together for the purpose of discussing not the future of the telephone, but its present.

Believing, however, as I do, that such a scheme will be the ultimate result of introducing the telephone to the public, I will impress upon you all the advisability of keeping this end in view, that all present arrangements of the telephone may be eventually realized in this grand system.

To achieve such a grand system would require telephones that would work over long distances because electronic amplifiers had not yet been developed. Bell's magneto telephone was unable to do that. There was just not enough power in sound waves to generate a strong electrical signal and provide the long-distance service that was envisioned. Although the magneto telephone had been tested successfully over a distance of 143 miles in December 1876, the transmission level was not robust and transmission was needed over thousands of miles, not just hundreds of miles.

The development of more powerful resistance transmitters that used batteries came soon with the application of a well-known phenomenon: contact resistance between two conductors is sensitive to pressure. Contemporary electricians knew that a binding post that is screwed down loosely caused a high resistance in the line. Similarly, a telegraph key had to be pressed down firmly to actuate the sounder properly. Thus a resistance that could change with the pressure from sound waves would modify the current from a battery, and the modified current would carry the characteristics of the sound wave.

Bell spoke his famous words "Mr. Watson—Come here—I want to see you" into a resistance transmitter on March 10, 1876, and they are frequently recognized as the first words transmitted by telephone (Bell 1876a, 40). Those words, however, were articulated into a liquid resistance transmitter that employed a wire dipped into an acid solution. A liquid transmitter was never used again, and its demonstration had no bearing on the first commercial telephones that were of Bell's magneto design.

Unlike Bell's magneto telephone, all of the resistance transmitters were technically unsophisticated, and initially their development fell into the realm of the tinkerer rather than the scientist. The first to gain practical traction was based on an 1877 idea by a dry-goods clerk, Emile Berliner, and was further developed by Francis Blake. Berliner is relatively well known for his invention of the gramophone flat disc. Berliner's patent diagram (fig. 4.1) shows the principle by which a metal diaphragm A vibrated and altered the pressure at its contact with a metal ball C, thus modulating the direct current from a battery (Berliner 1891). The

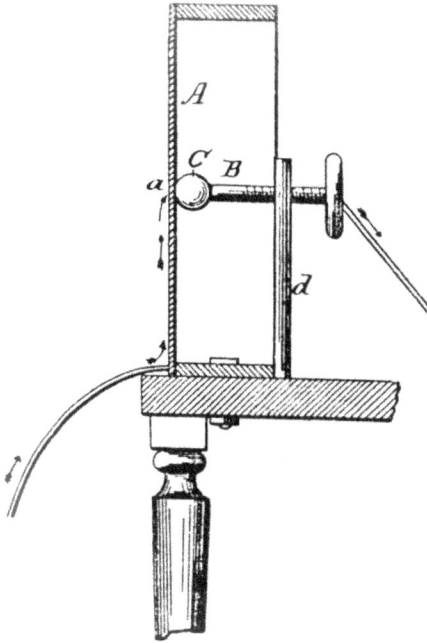

FIGURE 4.1 Drawing of a pressure-sensitive resistance transmitter from Berliner's patent. (Reprinted from Berliner 1891.)

result was an undulating current just like that from Bell's magneto telephone, only proportionally larger.

Blake's 1881 rendition of this was a rather complicated arrangement of leaf springs, platinum and carbon electrodes, and a coil (transformer) to pass only the alternating component of the undulating current to the line (Blake 1881). These parts were all mounted in a box that measured 5 inches by 6 inches (12.7 cm by 15.2 cm) by almost 3 inches (7.6 cm) deep (see fig. 5.1, left).

Although the Blake transmitter provided a stronger voice signal than Bell's magneto telephone, the Blake instrument had its limitations as well. With only one point of contact, the battery current had to be kept small or else the contact would overheat. In 1881 Henry Hunnings, a clergyman in England, patented his idea for a multiple contact powdered-carbon transmitter (fig. 4.2) (Hunnings 1881). The Bell Company purchased Hunnings's patents and redesigned the transmitter in 1885 with a horizontal diaphragm, calling it the long-distance transmitter (see fig. 5.1, right) (Fagen 1975, 73).

In 1890, Anthony White, a Bell Company engineer, developed the so-called solid-back transmitter (fig. 4.3) that would be typical of transmitters used by all

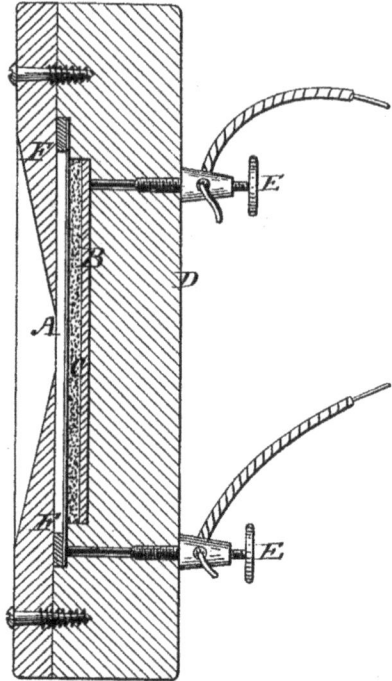

FIGURE 4.2 Drawing of a powdered-carbon transmitter from Hunnings's patent. (Reprinted from Hunnings 1881.)

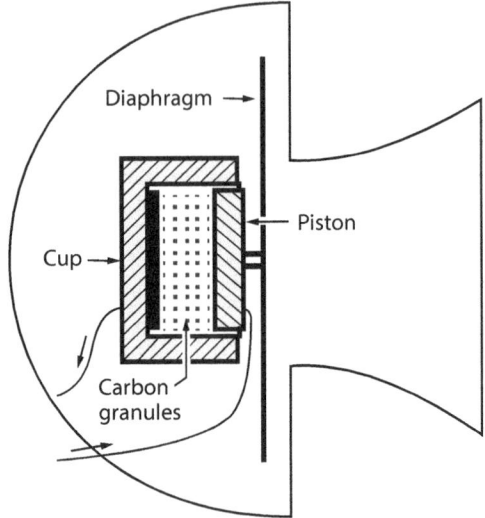

FIGURE 4.3 Diagram of White's 1890 solid-back transmitter.

manufacturers well into the 1930s (White 1892). Both the Blake transmitter and the long-distance transmitter were replaced by White's solid-back transmitter with an overall diameter of 3⅛ inches (7.9 cm).

But all of these resistance transmitters had their own problem; they had to be rigidly maintained in the position as shown. In the solid-back transmitter, the carbon particles had to be loosely packed and thus there was always some incomplete filling, with a gap at the top. If these transmitters were tilted away from their upright positions, particles would start to fall away from one of the electrodes. Thus the performance of these transmitters was position sensitive. Consequently, all telephone designs of this period were constrained to hold their transmitter in a fixed position. Handheld transmitters, such as in a handset, could not be used without sacrificing performance. Accordingly, the Bell Company refused to introduce a telephone with a handset until a position-independent transmitter was developed in the 1920s—and there was another reason, discussed in chapter 6.

5

IT'S ONLY
APPLIED DECORATION

The earliest resistance transmitter used by the Bell Company were the Blake transmitter of 1878 and the long-distance transmitter of 1885. They would be held upright on desk stands (fig. 5.1) and would be accompanied by a wall-mounted ringer box that housed the switch, ringer, and magneto, or they would be mounted rigidly on a box wall phone that contained the same components.

By the time the solid-back transmitter arrived, the switch hook and receiver were generally moved to the desk stand, and a smaller ringer box was used for signaling. The earliest stands had shafts of spun brass sheet with metal bases that were made with spun brass sheet over a cast-iron base for stability, but this construction was shortly replaced by solid brass parts (fig. 5.2). Brass tarnishes easily and becomes unattractive, so all early brass parts were nickel-plated.

Cabinet-style phones were popular and could be made to contain all the components in one place. The first such combination was a wall phone (fig. 5.3) made by Western Electric for the Bell Company and introduced in 1882 with a Blake transmitter (Fagen 1975, 130). Quart-size wet cell batteries were enclosed in the lower part of the cabinet, and the total assemblage was significantly larger than the Williams' Coffin sets.

Other similar stand-alone telephones were made during this period. Figure 5.4 shows a 6-foot-tall set from 1894 that sits on the floor like a grandfather clock.

FIGURE 5.1 Blake transmitter, 1878 (left), and long-distance transmitter, 1885 (right), on desk stands (Photos courtesy of the AT&T Archives and History Center.)

FIGURE 5.2 White's solid-back transmitter on desk stands, 1895 (left) and 1902 (right). (Photos by Matthew Gay.)

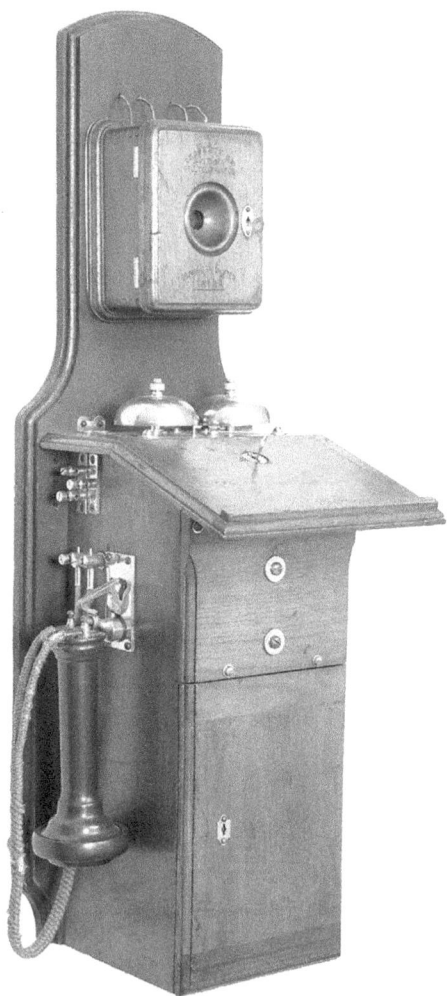

FIGURE 5.3 Wall-mounted telephone, 1882. (Photo by Matthew Gay.)

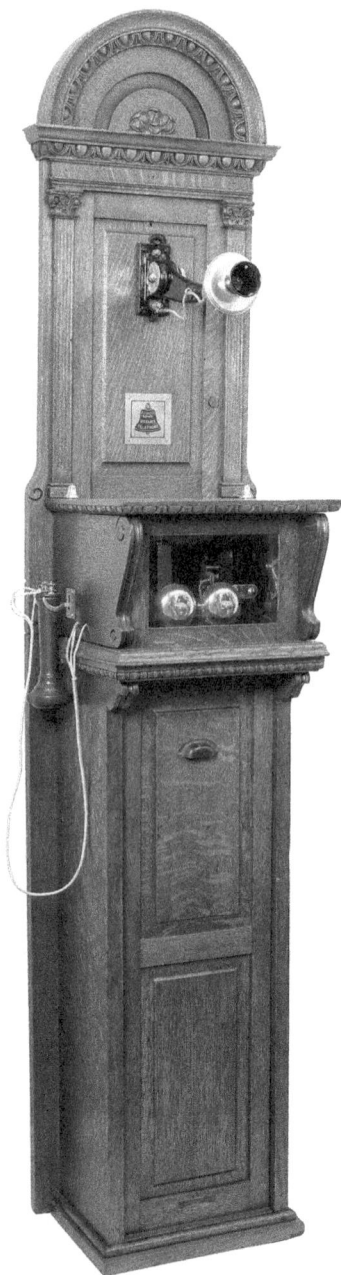

FIGURE 5.4 Floor-standing telephone, 1894. (Photo by Matthew Gay.)

A so-called vanity telephone (fig. 5.5) was built like an old school desk in a style introduced in 1895. No new convenience features had been added, and the only appearance enhancements were in the beautiful woodwork.

When dry cell batteries became available around 1900, smaller wall phones were manufactured (fig. 5.6). Notice in all cases that the transmitters are rigidly fixed in their upright position.

FIGURE 5.5 Vanity telephone, 1895.
(Photo by Matthew Gay.)

FIGURE 5.6 Wall telephone with dry cell batteries, 1907.
(Photo by Ted Mueller.)

Almost all of the telephones in service by the end of the 1800s were of the local battery type; that is, they all contained batteries for talking and magnetos for ringing. Although this type of phone would be used for another 50 years in some rural areas, no significant design changes were made other than to further simplify cabinets and share components with the central battery telephones that would come next.

6

CENTRAL BATTERIES AND THE BOOSTER CIRCUIT

E ven dry cell batteries were large and required changing periodically; magneto generators were also large and they were expensive. Hammond Hayes,
an engineer with the Bell Company, devised a method in 1892 that would
permit many telephones to operate off of batteries at a central office (telephone
exchange) rather than using local batteries in each telephone (Hayes 1892). Using this method plus other components at a central office, the magneto generator as well as the batteries could be eliminated from customers' telephone sets.
From left to right, figure 6.1 shows two dry cell batteries (two or three required),
a magneto generator, and a receiver for size comparison. The terms central battery and common battery were used interchangeably to describe this new system.

Without batteries and magnetos, wall sets (fig. 6.2) could be made smaller,
lighter, and less expensive than the local battery sets. For example, the local battery telephone in figure 5.6 weighed about 25 pounds (11.4 kg), compared with
9½ pounds (4.3 kg) for the corresponding central battery telephone in figure
6.2. Further, the nickel-plated desk stands (fig. 5.2) could still be used, although
with smaller subsets, which looked just like figure 6.2 without the transmitter
and receiver.

By the turn of that century, it was clear that central battery systems would become dominant, but simple central battery circuits in telephones were not nearly
as efficient as the standard local battery circuit (Meyer 2018, 128). To improve the

FIGURE 6.1 Size comparison of dry cell batteries, magneto, and receiver.

FIGURE 6.2 Western Electric's first wall phone with the booster circuit, ca. 1900. (Photo by Marketa Ebert.)

performance of central battery telephones, Charles Scribner, Western Electric's chief engineer, developed a new telephone circuit in 1897 that became known as the booster circuit (Scribner 1901).

The importance of the booster circuit cannot be overstated. In 1907 Theodore Vail, AT&T's long-time president and a friend of Gardiner Hubbard, re-affirmed the goal of universal service and established policies to reach that goal (AT&T 1908, 28; Brooks 1975, 68). Vail's universal service was an extension of Alexander Graham Bell's grand system. The concept involved local service in large and small communities—as well as in rural areas—with the local centers connected together by long lines such that any person could talk to any other person regardless of their locations. Thus long-distance operation was built into the concept, yet until around 1915 there was no practical means available to amplify telephone voice signals. Therefore, the signal voltage put on the line by an individual's telephone was all that was available for transmission through switches to the other end of the line, and the booster circuit provided a stronger signal.

The use of loading coils and amplification would come, but for the first years of the 20th century the only way to reduce the losses on long lines was to use copper wires of large diameter. With 165-mil (4 mm) wire, the booster circuit, resistance transmitters, and matched resonances of receivers and transmitters—all the tricks then available—the length over which conversation could be carried was not much more than 1,000 miles (Fagen 1975, 234). This was not enough to get universal coast-to-coast service, but it was better than any other telephone system in the world.

Although the booster circuit provided improved transmission, it had a unique problem of its own. Not only did the booster circuit put a powerful signal on the line, but it also put a powerful signal on the receiver of the transmitting phone (i.e., the sound heard by speakers of their own voice, which was called a sidetone). This was a circuit that only its mother could love. This large sidetone had a strong impact on design—contributing to a delay of about 25 years the introduction of handset telephones in the Bell System—as discussed in chapter 10.

7

SIMPLIFICATION AND
STANDARDIZATION

By 1900 there were about 2 million Bell telephones in service, and that number would double in a decade (AT&T 1907, 18). With that growth rate, the Bell System needed a more simplified product line and a massive new facility to keep up with manufacturing. Construction was thus begun in 1903 on the Hawthorne Works in Cicero, Illinois, just a few miles north of today's Chicago Midway Airport (Western Electric 1978). (For reference, the Wright Brothers' first flight was made in 1903 and Midway airport was built in 1926.) Hawthorne was so self-sufficient that it had its own hospital, fire department, power plant, foundry, machine shops of course, recreation facilities, and a night school for employees. It even had its own railroad to take supplies to its many loading docks and deliver products via connection with all the rail lines in Chicago. Called Manufacturers Junction Railway, it was only a few miles long, but was romanticized by Arthur Shilstone's watercolor collage Christmas card (fig. 7.1), commissioned by Western Electric (Polhemus 2015, 62).

Incidentally, Hawthorne became legendary for reasons other than manufacturing telephones and equipment. A major study of working conditions on productivity was conducted at the Hawthorne Works in the 1920s and 1930s by the Harvard Business School (Brooks 1975, 167). One result of the Hawthorne study, now known as the Hawthorne effect, showed that workers under study increase their productivity in spite of other factors.

FIGURE 7.1 Arthur Shilstone's watercolor of a Western Electric train at the Hawthorne Works. (Courtesy of the Arthur Shilstone family.)

For the first decade of the 20th century, the styles of the Bell System's central battery phones were simplified (fig. 7.2), but telephones continued to be made at older plants. Wood and nickel-plated brass were still the most often used materials, and the position-sensitive transmitter remained mounted rigidly. Other manufacturers made similar instruments for the growing central battery lines. Desk stands such as the one shown in figure 7.2 became referred to as candlestick desk stands, and they required a subset (fig. 7.2, right) to form a complete telephone.

FIGURE 7.2 Two typical central battery telephones of the 20th century's first decade. (Photos by Carroll Haugh.)

FIGURE 7.3 Two typical central battery telephones of the 20th century's second decade. (Photos by Ted Mueller and Carroll Haugh.)

By the second decade of the 20th century, similar telephones were being manufactured at the Hawthorne Works and wood was eliminated from their construction. No major design changes were made (fig. 7.3), but further simplification and cost reductions were achieved and all brass parts were then painted black (japanned). The black metal subsets perhaps gave rise to the generic term "black box" for a box containing mysterious components.

As evident above, design changes were still not being made for the user—changes that could be considered industrial design. All of the changes had been motivated by improved performance, ease of manufacture, and economy of operation. The same would be true for the next big innovation: the introduction of automatic dialing. In fact, many customers were reluctant to lose the friendly voice of a telephone operator and take responsibility themselves for making the correct connection. But cost incentives were overwhelming.

By the time automatic dialing was initiated by the Bell System in 1919—somewhat later than others—there were almost 200,000 telephone operators in the US, still using transmitters that were kept in a fixed position by a breastplate (fig. 7.4). And the number of telephone operators was growing rapidly (Daugherty 2022). This situation was not sustainable, and concepts for automatic dialing and switching had already been patented in 1879 by Daniel Connolly, Thomas Connolly, and Thomas McTighe (Connolly et al. 1879). The concept was engineered and put in place initially by the Strowger Company in La Porte, Indiana, in 1893 (Lindquist 1957). Strowger's company expanded, and in 1901 became the Automatic Electric Company, a competitor of Western Electric. But the Bell System

FIGURE 7.4 One of 200,000 Bell System telephone operators, ca. 1920. (*Disconnected* / Everett Collection / Shutterstock.com.)

did not introduce automatic switching during the early years of the 20th century and seemed instead to concentrate on universal service.

Automatic switching made further use of the direct current provided by batteries at the central office; this current would start to flow when a caller lifted the receiver off the switch hook. Prior to completing connection to a called party, the direct current at the central office passed through the coil of a relay's electromagnet, causing it to pull on an armature that was connected to a mechanical ratchet and pawl (or dog). Just as each swing of a clock pendulum advances the escape wheel by one tooth, each stroke of the relay advanced the ratchet wheel in the switch by one tooth with clockwork precision. In this manner, the desired sequence of switches would be selected. At the caller's end, a rotary dial was used to interrupt the current a number of times corresponding to the digit selected. Thus the rotary dial itself was just a switch.

When automatic switching was initiated by the Bell System in 1919, minor changes were made in their candlestick desk stand's design (fig. 7.5) to move the shaft off center and accommodate a dial (AT&T Archives 1992, 37). Other

FIGURE 7.5 Desk stand with off-center shaft to accommodate a dial, ca. 1920. (Photos by Carroll Haugh and Ted Mueller.)

manufacturers had done the same thing. Then, to further simplify production, so-called apparatus blanks were used when a dial was not needed for manual service. When such a blank was used in a candlestick desk stand, a cast-iron weight was added to compensate for the missing weight of the dial, thereby assisting in balancing the desk stand. Apparatus blanks continued to be used for all the later telephone designs through the postwar No. 500 telephone.

At about that time, a sliver of user preference crept into the customer-facing part of the dial. Overall, the size of the dial, size of the finger holes, and direction of rotation all had evolved during the 20-plus years that Automatic Electric had been using dials. But in 1918 William Blauvelt of the Bell System filed an application for a patent on a new numbering scheme that was then used for more than 40 years (Blauvelt 1922).

In growing metropolitan areas, customers might not be able to remember seven digits in long telephone numbers that would be needed (Bell Labs 1952). Studies had shown that five-digit and six-digit numbers could be remembered, but errors became numerous when trying to remember seven-digit numbers. The seventh digit seemed to be the straw that broke the camel's back. Blauvelt

FIGURE 7.6 Lettering scheme designed by Blauvelt for ease of remembering.

proposed naming the central offices (exchanges) and identifying a subscriber with that name plus a few numbers. A telephone exchange name and four or five digits could be easily remembered.

By adding letters to the numbers on a dial face (fig. 7.6), a combination of one or two letters and four or five numbers could be dialed, where the letters were chosen to suggest the exchange name. For example, a subscriber could dial PE6-5000 to reach Pennsylvania 6-5000 (the old Hotel Pennsylvania in New York City), although the number would be 736-5000 in the switch banks. If a manual exchange had not been converted to automatic dialing, a caller could still ask the operator for Pennsylvania six-five-oh-oh-oh. The letter Q was omitted to avoid confusion with the letter O, and Z was sometimes added to the number o. Letters from that earlier period are still present on today's cell phones.

PART 3

DESIGN BLOOMS

B y the third decade of 20th century, two major events had occurred that might have impacted telephone design: invention of the vacuum tube amplifier and World War I. (The impact of the Great Depression will be discussed in Part 5.) Neither of these two events had an appreciable effect on the design of the telephone itself, however.

In 1907 Lee de Forest filed an application for a patent on a three-element (triode) electron tube that was intended to provide a sensitive detector for radio receivers (de Forest 1908). Astonishingly, it would be five years before it was recognized that this electron tube could also be used as an audio amplifier (Fagen 1975, 260). In October 1912 the tube was used as an amplifier in a demonstration to Bell System officials, and H. D. Arnold of the Engineering Department at Western Electric began a study to improve the device for telephone use. In less than 12 months the Bell System had developed and tested an improved device (a high-vacuum electron tube) on a line between New York and Baltimore, and in January 1915 coast-to-coast telephone service began between New York and San Francisco using this amplifier.

Until that time, the Bell System had been using the booster circuit in its telephones to produce a strong signal for long-distance transmission. The large sidetone of this circuit could produce howling by acoustic feedback, and this shortcoming was one reason the Bell System delayed introducing a handset—a convenient handle that holds both a transmitter and a receiver. The Bell System could have used the vacuum tube amplifier and returned to a less powerful telephone circuit with reduced sidetone, but using amplification was not a simple matter for signals that usually went both directions over the same two wires. Thus amplifiers were used only on long lines where four wires were employed: one pair for signals going west, for example, and the other pair for signals going east. Telephones with robust signal production were still relied upon.

In 1917, just two years after the beginning of coast-to-coast long-distance service, the United States entered WWI. Demands for domestic telephone service soared and huge quantities of Bell System equipment were shipped overseas (Brooks 1975, 150). The number of Bell System telephones in service reached 10 million. For wartime efficiency, the Bell System was briefly taken over by the US government, but management was not replaced and business flourished. America's involvement in WWI was short, however, and the war did not appear to have an impact on engineering and design in the Bell System.

But what did have an impact on telephone design in the 1920s was the artistic interest of AT&T's revered president, Theodore Vail, and to understand Vail we have to go back in time.

8

THEODORE VAIL,
PATRON OF THE ARTS

E
arly in the 20th century, the Bell System openly embraced the arts. To
see how this course was set, one needs to look at the life of a single person, Theodore Vail, the long-serving president of AT&T. His life followed an arc from farming to art, and he took the Bell System with him. Vail's
interest in art would permeate the organization and lead to improved aesthetics in its telephones.

Theodore Newton Vail (1845–1920) was "one of the great figures in the Bell
System—a master of organization, a leader of limitless courage and resources,
unselfish and utterly honest in his management of the company's affairs" (Fagen 1975, 29). Born in Ohio, Vail spent most of his growing-up years in New Jersey. In 1866, with the Civil War over, Vail's family moved to a farm in Iowa, and
by March 1868, Vail told a friend, "I have had all of that dam' farm I want. I am
going where I can make some money" (Paine 1921, 36). With that, he took a job
with Union Pacific Railroad in Pine Bluffs, Wyoming, where the golden spike
was driven during his tenure—an area still populated by Native Americans.

Just a year later, Vail became tired of the Wild West, and he asked an influ-
ential uncle for help. An official letter subsequently arrived for the postmaster
in Pine Bluffs, and it said, "Find Vail and give this to him" (Paine 1921, 41). The
letter informed Vail that he had been given an appointment as a mail clerk,

which meant he would work for the US Post Office rather than the railroad. He would ride the Union Pacific trains with the mail between Pine Bluffs and Omaha—and he could live in Omaha. He continued in this job for more than three years, at which time he was asked to move to Washington to take a job as a special assistant to the general superintendent of the Railway Mail Service. When his boss retired in early 1876, Vail was given the job of general superintendent of the Railway Mail Service, a job that would serendipitously lead to something bigger.

Sometime between late 1877 and early 1878, a US congressional postal committee made an inspection tour of their postal railway service. Because of his position, Vail was put in charge of the tour that went all the way from Washington, DC, to California. Gardiner Hubbard, the sponsor of Alexander Graham Bell's early telephone work, was a traveling member of that committee, and during the long tour, Hubbard and Vail became close friends.

Hubbard had chartered the Bell Company in July of 1877 and needed someone to manage company affairs. Keep in mind that the telephone company was less than a year old and was facing many lawsuits over Bell's 1876 patent, so the future of the Bell Company was not at all certain. It was a risky move for Vail, who had a good position with the US Post Office, but in May 1878 Vail accepted the position of general manager of the Bell Company, which was soon reorganized with the following officers (Paine 1921, 123):

> Gardiner G. Hubbard, president
> Thomas Sanders, treasurer
> Alexander Graham Bell, electrician
> Thomas A. Watson, general superintendent
> Theodore N. Vail, general manager

Several reorganizations followed, but Vail remained as general manager and was located in Boston. In early 1885, a further reorganization produced the American Telephone & Telegraph Company, and Vail was elevated to its president (Fagen 1975, 34, 42). Although AT&T headquarters was incorporated in New York, Vail maintained his principal residence in Boston, and with his new affluence, he became a patron of the arts (Paine 1921, 147, 186–87).

By 1887, though, Vail had exhausted himself and made a number of bad personal investments. His health suffered, and he resigned his position. However, in 1907, Vail came out of retirement to again become president of AT&T. Under his second administration, research, development, and management would be consolidated in New York. Then, in his essay in the 1910 annual report, Vail mentions the arts—seemingly out of the blue (AT&T 1911, 27):

> The development of the arts, the necessity of extensive laboratories and experimental departments, with technical staffs competent to keep abreast of modern progress and find out how to utilize all of everything, the large gross production at small margin of profit, the large capital requirements necessary to conduct business on these lines; all these place modern industrial enterprises either beyond the financial ability of any one individual, or far beyond.

This commitment to the arts was astounding, and less than three years later AT&T began construction of an architecturally significant headquarters building at 195 Broadway. To top off this building, a statue was commissioned in 1914, sculpted by Evelyn Beatrice Longman, and hoisted to the roof in 1916 (AT&T Archives 1992, 36; Gray 2000). This towering figure, captured by photographer John Barrington Bailey in figure 8.1, became one of New York's largest sculptures, after the Statue of Liberty seen in the background (Harris 2006, 24). The gilded *Spirit of Communication* had earlier names but is most commonly referred to as *Golden Boy*.

Fostering the arts had thus become intrinsic to the Bell System, and the Bell System would eventually turn its attention to aesthetics in its telephone designs. But progress was being made on position-independent transmitters and better telephone circuits, so with success in sight, the edict went out on May 18, 1918, from George Thompson, head of the Mechanical Department, to begin a comprehensive program for the development of a user-friendly handset (Fagen 1975, 43, 144). The objectives were as follows:

1. Dimensions must be based on head measurements.
2. The handset must work properly in any position.
3. Howling must be avoided.
4. The handle and all exposed parts must be made of insulating material.
5. Carbon noise must be avoided.
6. The transmission performance must be equal to that of the most efficient transmitter and receiver in current use so that the handset could be interchangeable with existing station sets.

FIGURE 8.1 *Golden Boy* statue on the tower of AT&T building at 195 Broadway ca. 1965. (Photo by John Barrington Bailey. Reprinted from Harris 2006, 24.)

In these objectives we see industrial design bursting forth, with comfort, ease of use, and safety emphasized from the user's point of view. Of course, performance was not to be sacrificed because the Bell System was still in pursuit of universal service—the grand system. In the following chapters, these six objectives are addressed, but not necessarily in the above order. First, we will survey the landscape leading up to Thompson's decree.

9

THE "FRENCH PHONE" HAS A HANDSET

A telephone handset needed a receiver as well as a transmitter, and the handheld receiver of the past was too bulky for this purpose. Fortunately, a small receiver about the size of a watchcase (fig. 9.1) had been designed earlier for use with a headband for telephone operators. Here again was an example where the design had to be developed from the inside out, and not merely created as a mold into which the engineers would squeeze the mechanism. You couldn't squeeze a horseshoe magnet into a watch case—or could you?

FIGURE 9.1 Two early watchcase receivers showing different magnet geometries. (Photo by Ted Mueller.)

With the permanent magnet cleverly reshaped and shrunken, this became possible, and by the 1920s Bell Labs had developed new magnetic alloys that improved performance and facilitated smaller magnets (Legg 1939). An early version used a flat bar magnet that had been bent in a U-shape with a single coil (fig. 9.1, left), and later versions used a semicircular shaped magnet with two coils (fig. 9.1, right). This small receiver made it possible to take the next big step forward in telephone design with handset telephones.

The earliest US patent for such a handset was filed by Robert Brown in 1879 (Brown 1880). Although his handset was not widely used in America, Brown moved to France and it became popular in Europe (Fagen 1975, 140). Starting in 1902, Western Electric manufactured a series of special purpose handsets for the Bell System, one of which is shown in figure 9.2. The transmitter in the Western Electric handset can be seen to be the same as the one used on the candlestick and wall phones shown in previous chapters, and the receiver was of the watch-case design.

In America, the Kellogg Switchboard & Supply Company introduced a handset desk stand called a Grabaphone in 1905 (fig. 9.3). The transmitter and receiver were fastened to a tubular handle in the Kellogg handset.

In Europe, Ericsson in Sweden (fig. 9.4) and Siemens & Halske in Germany (fig. 9.5) were also producing similar telephones, often called "French phones," and all of these were being marketed in America.

Sally Clarke argues that the Bell System was pressured by competition to introduce a handset telephone similar to the so-called French phones (Clarke 1998, 166). For that phone's popularity, Clarke cites the experience of returning veterans from World War I, a 1921 movie that glamorized the phone, a 1926 sales ad by Mor-Tel Corporation for such a phone,

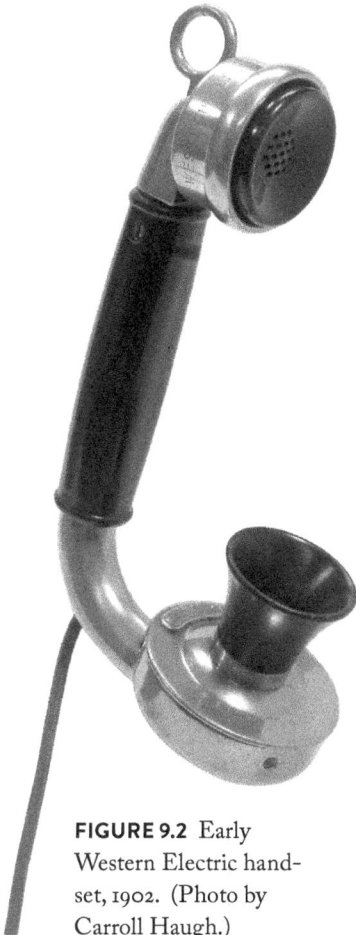

FIGURE 9.2 Early Western Electric handset, 1902. (Photo by Carroll Haugh.)

FIGURE 9.3 Kellogg Grabaphone handset telephone, 1905. (Photo by Carroll Haugh.)

FIGURE 9.4 Ericsson handset telephone, 1919. (Photo by Truls Nord, courtesy of Tekniska museet.)

FIGURE 9.5 Siemens & Halske handset telephone, 1919. (Photo by Martin Feuz.)

and a Sanka coffee ad from 1927 that elegantly featured a French phone. Although there is no question that the French phone enjoyed a certain popularity, the argument that this popularity pressured the Bell System to develop a handset telephone is unpersuasive; all of the incidences cited occurred after the Bell System had launched an aggressive effort to develop such a telephone on May 18, 1918.

There is nevertheless no denying that the Bell System fought to prevent these other phones from being connected to their system. While a subscriber in Manhattan might have been satisfied with their French phone's low performance for local calls, a caller from Upstate New York, for example, would undoubtedly—and mistakenly—blame the telephone company for a poor connection to Manhattan. Having produced several different handsets around the turn of the century, the Bell System was no stranger to their convenience but had resisted widespread introduction for practical reasons.

10

PROBLEMS SOLVED FOR A NEW HANDSET

OBJECTIVE 6: TRANSMISSION PERFORMANCE

The Bell System's commitment to universal service had led to the powerful booster circuit, and the company had made a deal with the devil with this circuit. Not only did the booster circuit put a powerful signal on the line, but it also put a large sidetone on the receiver of the transmitting phone. This sidetone was noticeable, it was annoying, it caused the user to speak softly, and it could produce howling by acoustic feedback from the receiver to the transmitter through a hollow handle. With old-style candlestick desk stands, these were not significant problems because the separate receiver could be moved while talking, and sound vibrations could not be transmitted through a handle to cause howling.

By 1918, George Campbell, head of AT&T's Electrical Department and famous for his circuit analysis, had shown mathematically that a simple high-gain anti-sidetone circuit (also a booster-type circuit) could be designed with a very small (theoretically zero) sidetone (Campbell 1937, 119). Although the new anti-sidetone circuit would not be ready until 1930, luck was with the Bell System. They were able to introduce an efficient handset telephone that still used the old booster circuit by utilizing a massive solid handle for the handset rather than a hollow tube.

OBJECTIVE 4: INSULATING MATERIALS

Using insulating material for the handset was not necessarily a problem. Because of the potential for electric shocks, this was a safety issue that had already been addressed in earlier styles. Notice that exposed terminals on early receivers (see fig. 7.2) had been eliminated in the following decade (see fig. 7.3). However, this objective would require the use of a material such as hard rubber or Bakelite for the entire handset structure, and the material must be proven to be strong enough to endure rugged service.

FIGURE 10.1 Drawing from Thompson and Harper's design patent. (Reprinted from Thompson and Harper 1922.)

Preliminary work on a new handset began almost immediately. On December 8, 1919, Walter Kiesel filed an application for a patent on a handset with a solid handle made of hard rubber or Bakelite (Kiesel 1922). The main purpose of Kiesel's work was to show that conducting strips could be embedded in the handle such that the required electrical conductors could also serve as reinforcement—like rebar in concrete—and thus strengthen the handle. Hollow tubes of those nonconducting materials, with the usual cord conductors, would not be rugged enough for severe usage in, for example, frontline military operations.

Work on the shape of the handset (fig. 10.1) was begun by George Thompson and Alfred Harper, who filed a design patent application on October 15, 1920 (Thompson and Harper 1922).

OBJECTIVE 3: NO HOWLING

Howling and sidetone had not been problems for the Kellogg and the European handset telephones because they did not use the booster circuit, which was patented and not available to them. A simple central battery circuit was used in Europe and it did not have a large sidetone, but it also did not put a powerful voice signal on the line. Kellogg used a different circuit in its

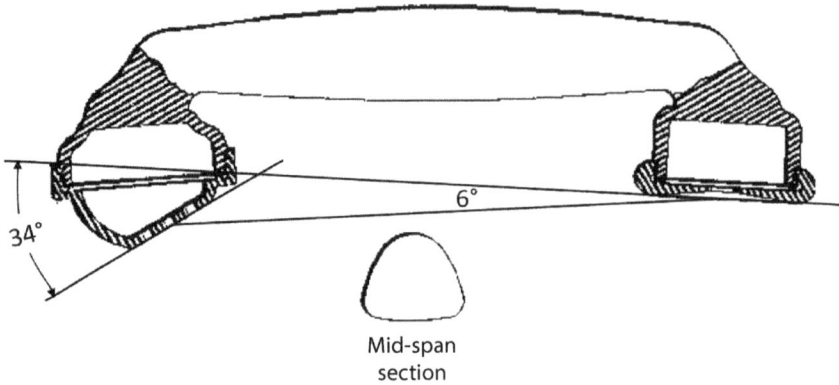

34°

6°

Mid-span
section

FIGURE 10.2 Drawings (edited) from Curtis's patent.
(Adapted from Curtis 1925.)

handset desk stand, but that circuit also suffered power losses from the transmitter to the line. Whereas in the booster circuit the sidetone and line voltages were greater (by about 65 percent) than the transmitter's output voltage, the sidetone and line voltages were lower than the transmitter's voltage (by about 25 percent and 40 percent, respectively) in the simpler central battery circuit (Meyer 2018, 130). On the best long-distance line of the time, with a line loss of 0.033 decibels per mile, this difference corresponded to about 8 decibels and a transmission distance of roughly 250 miles (Fagen 1975, 323). This was a performance penalty the Bell System refused to accept.

On December 24, 1920, Alfred Curtis filed for a detailed patent on Thompson and Harper's design, or a very similar one. The main purpose of Curtis's work was to ensure that howling could not be caused by acoustic feedback through the handle (Curtis 1925). By using a massive handle of phenolic plastic (Bakelite), the handle's resonant frequencies were kept below 600 cycles per second and therefore below the range of significant acoustic feedback from the receiver to the transmitter. Curtis's patent diagram is shown in figure 10.2. In this and the following three figures, numerical parts identifiers have been digitally removed (edited) and construction lines added to show important angles and other features that are discussed below. Notice that the triangular cross section of the handle, which was implied in Thompson and Harper's design, is shown explicitly in Curtis's patent.

OBJECTIVES 2 AND 5: POSITION INDEPENDENT AND NO CARBON NOISE

On December 30, 1922, Harold Dodge submitted a patent application for a further evolution of this handset (Dodge 1928). Although the ostensive purpose of Dodge's patent was to show improved sound articulation with a new transmitter mouthpiece, the altered outline of the handset reveals different angles and contours (fig. 10.3). Dodge appeared to be incorporating results from head measurements that were emerging (see chapter 11). On December 31, 1923, George Thompson filed an application for a design patent on the shape of the handset depicted by Dodge (Thompson 1925).

By the way, Harold French Dodge (1893–1976), whose name comes up again below, became well known in the field of quality assurance (American Society for Quality n.d.). He was a fellow and founding member of the American Society for Quality and did pioneering work on data sampling. Dodge had degrees in engineering, physics, and mathematics from MIT and Columbia, and after retirement from Bell Labs held a position as professor of applied mathematical statistics at Rutgers University.

A few months after Dodge submitted his patent application, on February 7, 1923, Harry Clarke filed for a detailed patent on a handset similar to that shown in Dodge's patent, and Clarke's patent provided both receiver and transmitter

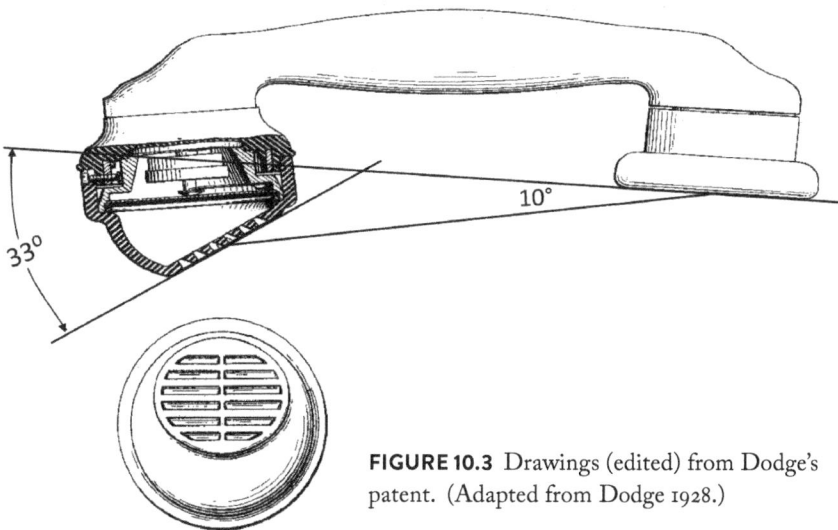

FIGURE 10.3 Drawings (edited) from Dodge's patent. (Adapted from Dodge 1928.)

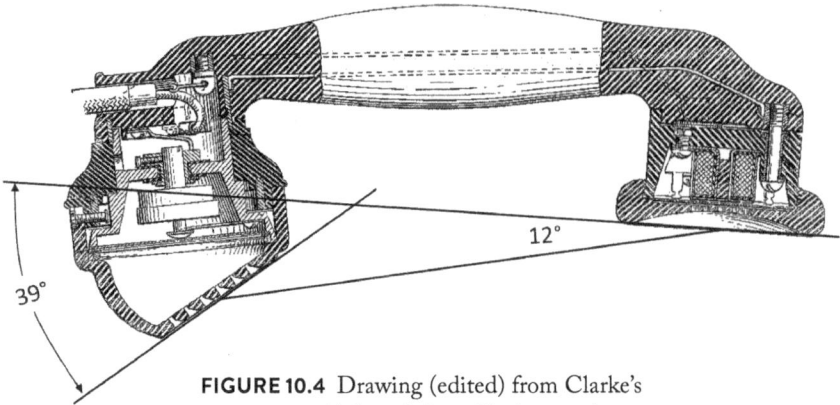

FIGURE 10.4 Drawing (edited) from Clarke's patent. (Adapted from Clarke 1926.)

details (Clarke 1926). The transmitter (fig. 10.4) appears to be Charles Moore's original position-independent transmitter of 1921 (Moore 1925). As hoped, it was confirmed that achieving positional independence also ensured that contact pressures and resistances in the handset transmitter were substantially independent of angular position, such that the change in carbon noise with position was practically negligible (Jones and Inglis 1932, 253).

The handset in figure 10.4 was actually produced and called the Type A handset. In 1924, "a sample of 490 units of the Type A design was manufactured and placed in service with Bell System employees at the American Telephone and Telegraph and Western Electric locations in New York City. Several deficiencies were found in this trial and redesign continued, passing through Types B, C, and D" (Fagen 1975, 150).

On August 20, 1925, Arthur Bennett and Charles Moore filed an application for a patent on the near-final form of the Type E handset (fig. 10.5), incorporating an improved version of Moore's barrier-button transmitter (Bennett and Moore 1929). Yet another improvement would be made in Moore's transmitter before incorporation in the eventual Type E handset.

The new position-independent transmitter became available before the anti-sidetone circuit was ready, and the new transmitter had a more uniform (flatter) frequency response than the old transmitter, which exhibited a strong output resonance around 1,000 cycles per second (Fagen 1975, 150; Jones and Inglis 1932, 255; Meyer 2018, 16–18). Based on testing, the Bell System found that elimination of the old resonance reduced the sidetone, thus causing the user to speak

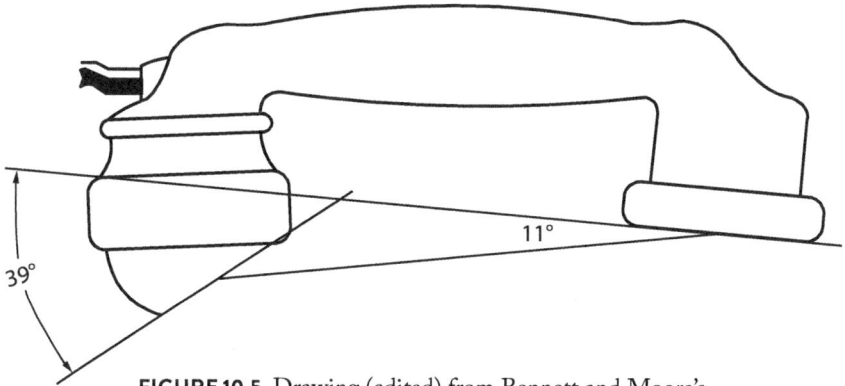

FIGURE 10.5 Drawing (edited) from Bennett and Moore's patent. (Adapted from Bennett and Moore 1929.)

louder and compensate for the overall reduction in transmitter performance. The reduced sidetone in combination with a massive phenolic plastic (Bakelite) handle also eliminated accoustic feedback through the handle. Thus the Bell System had solved the problems sufficiently to introduce the new handset in 1927 with the old booster circuit before the anti-sidetone circuit became available in 1930. This good fortune also permitted the Bell System to get further use out of old sidetone subsets, since they were separate components.

11

HANDSET DIMENSIONS

T he first of George Thompson's objectives, which are listed at the end of chapter 8, was that dimensions of the handset must be based on head measurements. Such anthropometric measurements had never been made before, so this was a pioneering—and massive—effort.

OBJECTIVE 1: DIMENSIONS BASED ON HEAD MEASUREMENTS

For comfort and maximum sound efficiency, there was a need to optimize the relative positions of the transmitter and receiver based on head measurements. Thus the Bell System, in parallel with the developments described in chapter 10, launched an anthropometric study to obtain important, or modal, dimensions for their new handset (Jones and Inglis 1932, 262). For this study, about 4,000 measurements of head sizes were made using a special gauge (fig. 11.1).

The gauge was built on a bar that had a receiver cap attached to it, and the distance along the bar was calibrated in centimeters. A slider moved along the bar and could be adjusted such that a scale attached to the slider touched the lips of the subject. This scale was also calibrated. A calibrated pin went through the bar and could be placed in contact with the cheek. Figure 11.2 is a diagram of

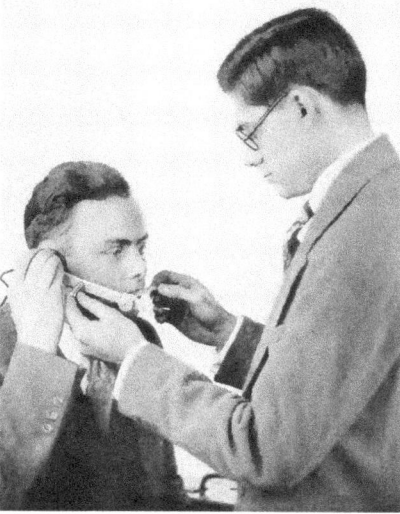

FIGURE 11.1 Head-measuring gauge in use by the Bell System. (Photo courtesy of the AT&T Archives and History Center.)

this gauge (minus the cheek-clearance pin) and shows a right triangle, of which the base and height were measured. The included angle was known (90 degrees), so applying a little trigonometry (side-angle-side) gave the remaining side and angles, which included the modal length (often called delta, δ) and the modal angle (alpha, α) that were of interest.

Results from all of the head measurements are shown in figure 11.3. The location of a mouth relative to the center of the receiver cap is given by its modal length and modal angle. In the large array of data on the right, distances are grouped in half-centimeter intervals and angles are grouped in 2-degree intervals. For example, 17 people had mouth locations that were about 15.5 cm from the center of the receiver cap at angles of 19 and 20 degrees. There are a total of 3,888 measurements in this display.

With reference to figure 11.3, it can be seen that only 22 measurements were equal to, or exceeded, 16 cm from the center of the receiver cap to the center of the mouth and would thus have resulted in interference; these long heads comprised just 0.6 percent of the total. In order to show the location of the majority

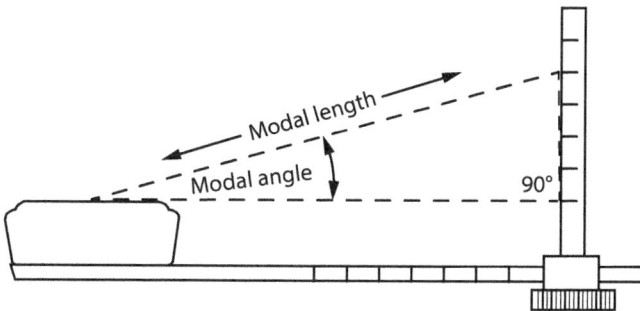

FIGURE 11.2 Diagram of the Bell System head-measuring gauge.

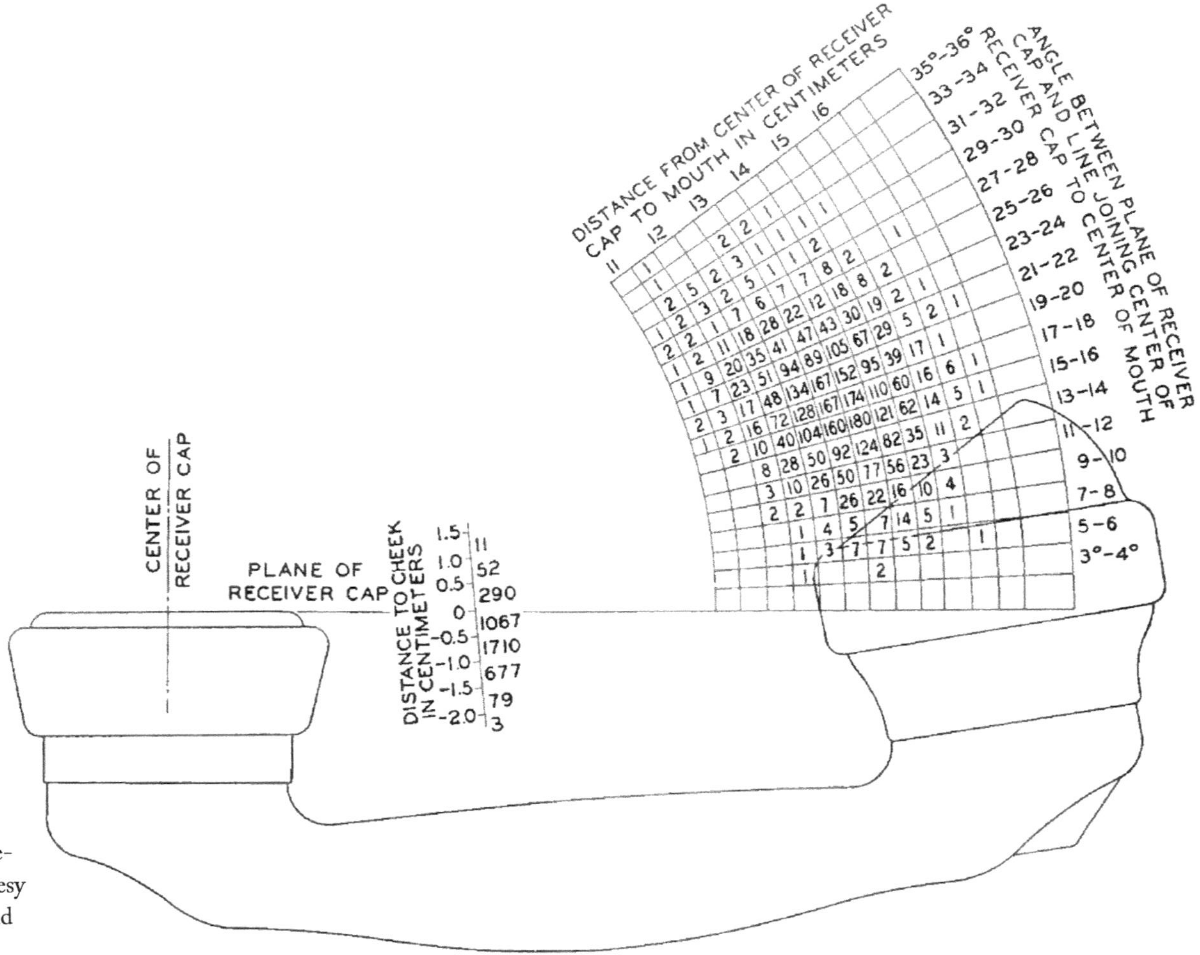

ANGLE BETWEEN PLANE OF RECEIVER CAP AND LINE JOINING CENTER OF RECEIVER CAP TO CENTER OF MOUTH

DISTANCE FROM MOUTH CENTER IN CENTIMETERS

	11	12	13	14	15	16	
35-36°							
33-34				2			
31-32		2	2	2	1		1
29-30		2	3	1	2	1	1
27-28	2	5	7	8	9	2	1
25-26	1	7	18	8	5		
23-24	2	11	18	47	43	30	29	5
21-22	2	2	11	28	47	67	17
19-20	9	20	35	41	89	105	52	95	39	16	6
17-18	1	7	23	51	48	34	61	74	110	60	16	5
15-16	3	17	28	67	80	121	62	14	11
13-14	2	6	40	104	160	124	82	35	2
11-12	2	10	26	50	92	77	56	23	3
9-10	8	28	50	22	16	10	4
7-8	3	10	7	7	14	5	1
5-6	2	2	4	5	1	3	7	7	5	2
3°-4°	1	1	3	1	7	2	2

DISTANCE TO CHEEK IN CENTIMETERS

1.5	11
1.0	52
0.5	290
0	1067
-0.5	1710
-1.0	677
-1.5	79
-2.0	3

PLANE OF RECEIVER CAP

CENTER OF RECEIVER CAP

FIGURE 11.3 Distribution of Bell Labs head measurements. (Illustration courtesy of the AT&T Archives and History Center.)

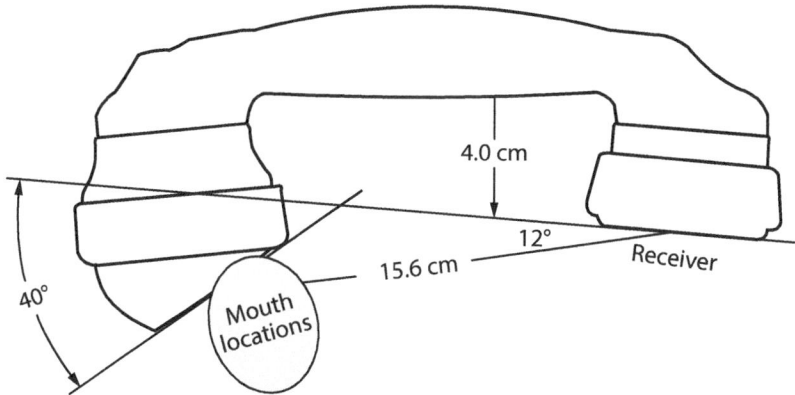

FIGURE 11.4 Outline and modal dimensions of the Type E handset.

of mouths in relation to various handset designs, we chose a region bounded by 12 and 15.5 cm (modal length) and 8 and 26 degrees (modal angle). More than 90 percent of the mouths are located within this region, which is shown in relation to the Type E handset in figure 11.4. Figure 11.3 is visually consistent with figure 11.4, which shows dimensions we measured on a production Type E handset.

Face (cheek) clearance was ensured by extending the transmitter and receiver from the handle by an amount that was greater than the distance-to-cheek measurements (–2 cm for the widest heads) plus an allowance for the thickness of fingers wrapped around the handle (not measured in the study, but about 1.3 cm). The measured clearance for the Type E handset was 4.0 cm; the needed extension (about 3.3 cm) was exceeded by the large size of early transmitter and receiver units.

Finally, there was one other angle that had to be set, and this was the tilt of the transmitter. This angle was not dictated by the anthropometric study results, and it would not be possible to orient the transmitter to face directly toward the mouth for all head sizes with a single fixed angle. By subjectively setting this angle at approximately 40 degrees, the transmitter would be at the side of the mouth, yet angled toward the mouth, for all head sizes.

If we look at the evolution of the Type E handset, we see that the distinctive shape of the handle appeared first in the 1922 patent design of Harold Dodge. A modal angle close to 12 degrees appeared for the first time in Dodge's design. Although we cannot measure the modal distance between the transmitter and receiver in these patent drawings, we can estimate that this distance in Dodge's design was about 15.5 cm based on relative shapes compared with our measurements on the production Type E. The modal distance appeared longer in the previous design. Thus we conclude that the critical or modal dimensions resulting from the anthropometric study were not present in Alfred Curtis's design, but appeared first in Dodge's design and were present thereafter.

In the Bell System, design of the handset had consumed the lion's share of work on their new telephone, so we have focused attention on handset designs. We found interesting trends and implications from measuring handset dimensions for the major designs that are featured in later chapters of this book. A summary of our measurements on pre–World War II handsets is presented in table 11.1, where values for the Western Electric Type E handset are shown in boldface for ease of comparison.

Table 11.1 Summary of pre-WWII handset measurements made by authors*

HANDSET IDENTIFICATION	INTRODUCTION YEAR	LENGTH, δ (CM)	ANGLE, α (DEGREES)	TILT (DEGREES)	CLEARANCE (CM)
Western Electric Type A	**1924**	**15.4**	**12**	**40**	**3.5**
Western Electric Type E	**1927**	**15.6**	**12**	**40**	**4.0**
Western Electric Type F	1937	16.2	11	41	3.3
Automatic Electric No. 1	1925	16.6	8	35	3.3
Automatic Electric Types 34	1934	16.0	17	58	2.8
Kellogg No. 730	1930	15.3	16	57	3.2
Kellogg No. 1000	1946	16.0	13½	33½	3.0
Stromberg–Carlson No. 1178	1929	14.6	13½	57½	4.4
Siemens & Halske Model 26	1928	16.5	8	50	1.9
H. Fuld & Company Frankfurt	1929	16.4	8½	51	3.0
Siemens Brothers Neophone	1929	15.0	13	63	2.9
L.M. Ericsson DE 702	1931	15.8	15	58	2.5
BTMC Antwerp	1935	16.0	12½	58½	2.5

*Approximate accuracy +0.1 cm, −0.3 cm, ±1 degree.

PART 4

DESIGN'S GROWING PAINS

Thompson's six objectives for a new handset did not include appearance and, in fact, few if any considerations were given to it. It turns out that the new handset was such an improvement over earlier devices that the handset's appearance did not become an issue. However, appearance of the mounting on which the handset was placed, a desk stand, became problematic.

12

COMPLAINTS
ABOUT APPEARANCE

D
evelopment of a desk stand (Western Electric called them handset mountings) for the new handset trailed the handset development. On December 28, 1922, George Thompson submitted an application for a design patent that was primarily about a new cradle (fig. 12.1), which was simply mounted on a shortened candlestick base (Thompson 1924a). On the same date, Thompson also submitted a detailed functional patent, so it is clear that Thompson started working on the cradle design much earlier than December 1922 (Thompson 1924b). This handset desk stand was referred to as the A type.

Although this desk stand was used in field trials with the Type A handset in 1924, production did not begin on the Type E handset until late 1926 at the Hawthorne plant. The Type E handset was then made available for more general use on the A-type desk stand in early 1927 (Fagen 1975, 151).

The handset design appears to have been finalized at that time, but the new cradle appears to have been introduced as a prototype on the candlestick base. A survey of collectors in two clubs during 2018 identified 14 A-type desk stands with D76869 stamped on the shaft rather than the expected A1 designation, and only one special set (different switch configuration) was marked with an A3 designation. D-numbers were drawing numbers and were usually put on prototype sets, apparently confirming the prototypic nature of these first handset telephones.

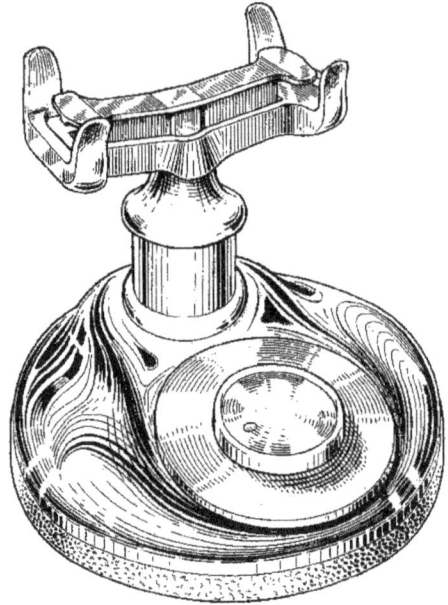

FIGURE 12.1 Drawing from
Thompson's design patent.
(Reprinted from Thompson 1924a.)

The A-type desk stand was short-lived because a problem was discovered with the cradle. Figure 12.2 shows an A-type cradle with the receiver hung up on one of the cradle prongs and thus not depressing the switch adequately. This receiver-off-hook, or busy, condition tied up valuable lines and was difficult for the telephone company to correct quickly.

The cradle design was soon changed, along with the construction of the base, and a revised B-type desk stand (fig. 12.3) was in service by December 1927 (AT&T 1927, 1). In the meantime, A-type desk sets were assigned to manual service, where the operator could more readily detect a problem. With dial service, a busy line could persist until a subscriber complained. A detailed patent application for the B-type desk stand had been filed in November 1927 by William Scharringhausen, and the main purpose was stated to be "to prevent false operation due to the handset failing to seat itself properly when placed on the mounting" (Scharringhausen 1931). Although not discernable in figure 12.3, the shape and spacing of the four prongs of the cradle were changed such that it was not possible to prop the handset up on a cradle prong.

Sally Clarke, who described the popularity of the so-called French phone, also described a letter dated January 24, 1927, that complained about the appearance of the A-type base (Clarke 1998, 177). In that letter, a subscriber was said to

FIGURE 12.2 Western Electric Type E handset and A-type base of early 1927 with receiver-off-hook error.

FIGURE 12.3 Western Electric Type E handset and B-type base of late 1927. (Photo by Carroll Haugh.)

be unhappy with the new handset telephone because it wasn't attractive like a French phone and described the Western Electric product as "being something awful." The letter was sent to Charles DuBois, president of Western Electric, and passed along to Frank Jewett, president of Bell Labs and vice president of AT&T.

The B-type desk stand had not come out yet when that letter was sent in early 1927. And the A-type desk stand was known to be a prototype. Nevertheless, and perhaps partly because of the letter, a nationwide review was conducted within the Bell System to determine the adequacy of the appearance of subscribers' equipment (Colpitts 1928, 1). The B-type desk stand had been introduced in late 1927, so the review focused on that desk stand and found it wanting.

The B-type desk stand was a result of engineers working without designers. It looks unbalanced and a bit tacky with its surface-mounted dial. After the nationwide review had been completed, instructions were given by AT&T's assistant vice president Edwin Colpitts to Bell Labs to carry out a development program "as promptly as possible" to replace the B-type desk stand with something more attractive (Colpitts 1928, 2).

13

SMALL CHANGES MAKE
A BIG DIFFERENCE

dwin Colpitts's 1928 letter to Bell Labs went so far as to specify that the new mounting would contain only a switch hook, a dial, and in some cases keys (line switches) and a related buzzer. As with earlier telephones, a separate subset would be provided for the primary ringer, induction coil, and condenser. Small sketches of the handset mounting and subset were provided in an attachment to Colpitts's letter, and the sketch for the handset mounting showed an oval base.

Less than three months later, Colpitts wrote back and gave approval of the design represented by a model that Bell Labs had given him (Colpitts 1929a). A photograph of what we believe is that model was included in a speech by AT&T executives in Tokyo in October of that year (Gherardi and Jewett 1930, 10). It can be seen to be a model by the absence of a pronounced seam between the cradle and the base—a seam that is present in the production set—and the model would have been made in the Bell Labs Model Shop on West Street, New York (Bell Labs 1925). A photograph of a fully functioning prototype of the D-type base, bearing the date "10-6-29" on the front edge, was shown in Kempster Miller's well-known book *Telephone Theory and Practice*, so it is clear that this design was completed by early 1929 (Miller 1933, 106).

The new D-type desk stand (fig. 13.1) had an elliptical base, a recessed dial, and improved lines that Colpitts cited as "important factors contributing to the

FIGURE 13.1 Western Electric Type E handset and D-type base
for No. 202 telephone, 1930. (Photo by Carroll Haugh.)

pleasing effect" (Colpitts 1929a). The handset was unchanged. Taken together, the Type E handset, the D-type desk stand, and the redesigned subset with its new anti-sidetone circuit comprised the No. 202 telephone, introduced in 1930.

Colpitts's letter of late 1928 also included the redesigned subset in the development program. It was considerably smaller than earlier subsets and was fitted with a Bakelite cover. According to Colpitts's sketch, the Bakelite cover was to be attached with two screws rather than a hinge as in older subsets. This detail plus the oval shape in the sketch of the new handset mounting suggests that the concepts for these designs were already in existence by the time Colpitts wrote his letter in 1928.

Credit for rendering the revised design that became the D-type desk stand is not clear, but we believe it belongs to Harold Dodge. In 1931, Dodge published a paper on squares and rectangles in which he pointed out the frequent reference in art history to a "golden section" whenever form and proportion were discussed (Dodge 1931). Properties of the golden section can be defined with the help of figure 13.2.

A rectangle or ellipse is said to be golden when the ratio of the minor dimension A to the major dimension B is the same as the ratio of the major dimension to the sum of both dimensions: $A/B = B/(A + B)$. This reduces to a quadratic

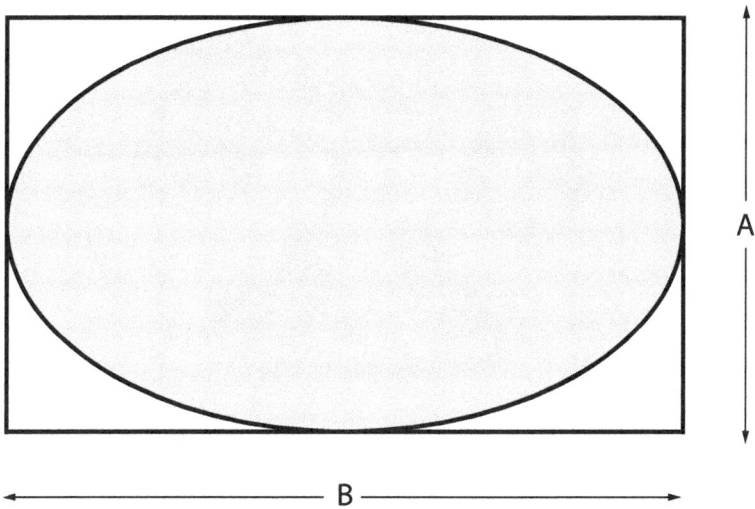

FIGURE 13.2 Ellipse inscribed in a golden rectangle.

equation that has a positive solution $A/B = \frac{1}{2}\sqrt{5} - \frac{1}{2}$, which equals 0.618 to three decimal places. Dodge appears to have been well acquainted with this ratio and thus had some artistic knowledge.

Dodge had been in the Engineering Department of Western Electric from 1917 until he shifted to Bell Labs during the company's restructuring in 1925. The head of Western Electric's Engineering Department as of 1905 (and presumably later) was George Thompson, whom we credit with the basic design of the handset desk stand (Fagen 1975, 43). Dodge had also been associated with the Research Department in the development of telephone instruments and allied products (Bell Labs 1931). So Dodge was in the right place in the company at the right time to have done this design revision.

Dodge also played an explicit role in the development of the handset with his patent for the transmitter mouthpiece. Although this mouthpiece was not used in the final design, remember that the final contours of the handset, which incorporated results from the anthropometric tests, appeared for the first time in Dodge's patent. While no claim was made in the patent description for these shape changes, they were profound because of the head measurements—and they did appear in Dodge's patent diagram. Because of his data-sampling and statistics expertise, it seems almost certain that Dodge would have been involved with the anthropometric study that took head measurements on the large sample of people and led to the handset parameters. By the way, information

FIGURE 13.3 Elliptically circumscribed view (border added) of Western Electric Type E handset and D-type base. (Photo courtesy of the AT&T Archives and History Center.)

on developments in the Bell System's earlier engineering departments was not readily available prior to the beginning of publication of the Bell Laboratories Record in 1925.

Finally, Dodge's coworker, Marion Dilts, also published a paper on the golden section in the same journal as Dodge's article (Dilts 1931). Dilts used a golden ellipse to show that the shape of the Type E handset on the D-type desk stand had pleasing proportions; the picture she used is shown in figure 13.3 with an outline added for emphasis. Notice that Dilts's golden ellipse is an imaginary outline of the whole telephone and does not refer to its elliptical base, which has dimensions 4 inches by 5 inches (10.2 cm by 12.7 cm) with an aspect ratio of 0.80 (not golden). So here was an article that cast this desk set as an attractive object, rather than an ugly one that was subjected to criticism in its earlier forms. It seems likely that Dilts would have selected this desk set as her sole example for the paper if her senior coworker actually did the final design modifications.

One later design detail of the D-type desk stand involved the material used to cover the base plate. On the predecessor candlestick desk stands, the base plate was covered with wool felt, and the candlestick phone could be dialed without its sliding across a smooth surface. The base plates were also covered with wool felt on the A-type and B-type desk stands, and those two were marginal with regard to sliding while dialing. Keep in mind that the frictional force that resists sliding is proportional to the telephone's weight, and the proportionality constant is the coefficient of friction of the fabric. So if we consider the weight of these three bases, without their receiver or handset, they are 3 pounds, 8 ounces (1.6 kg) for the brass candlestick; 2 pounds, 5 ounces (1.0 kg) for the brass A type; and 2 pounds, 10 ounces (1.2 kg) for the steel B type.

When the D-type base first came out, however, it was made of cast aluminum and weighed barely 2 pounds (0.9 kg). It was so light that, if fitted with a felt-covered base, it had to be held firmly with one hand while dialing with the other. (How did one hold the handset?) Consequently, a more expensive suede covering was chosen for the base because it had a higher friction coefficient (Western Electric 1935a, 70). This change was sufficient to prevent sliding. Aluminum had other disadvantages, such as thread wear in the bosses where the base plate was attached and a propensity for pitting with age. Later in the production run, the material for the D-type base was changed to a zinc alloy at 3 pounds (1.4 kg) without handset, and a zinc alloy would also be used in the next generation of Bell telephones.

A final design detail of the Type E handset was added late in its production run. The handle and the end caps were all formed in two-piece molds that left a parting line or small ridge of expelled Bakelite where the molds came together. This was always at the locally widest point, and the unsightly parting line required careful grinding and buffing of a large area to produce the necessary surface smoothness (Huxham 1939, 214). The development of an ingenious method of removing the line automatically by machining a groove at the parting line was an improvement from the standpoints of both appearance and economy. These grooves were attractive and were not disguised in subsequent handsets designed for the Bell System.

Looking back at the evolution of the design of the handset and the desk stand, together referred to as the Western Electric No. 202 telephone in its final version, it appears that Thompson deserves credit as its architect. However, with both the handset and the base, significant appearance changes were made to Thompson's design. Although speculative, it is likely that Dodge provided these touches.

With the arrival of the handset, wall phones that held the transmitter in its upright position became less important; it is easier to move the handset to the mouth than to move the mouth to a phone. No companion wall set was made for the No. 202, although a wall-mounted bracket (C-type) for the handset was made, and that combination is sometimes called a space saver.

14

DESIGN MISSTEP IN
THE BELL SYSTEM

A T&T headquarters, which was well aware of complaints about the B-type desk stand, apparently did not realize that Bell Labs already had talented designers within its ranks who had the artistic capabilities to produce attractive designs. Therefore, as the D-type design was being developed in-house, Bell Labs was directed to launch a parallel effort in the spring of 1929 to enlist entirely new designs from prominent artists (King 1932, 1).

Why would such a misdirected project be launched? Who at headquarters had the authority to force the labs to do it? After all, Frank Jewett, the president of Bell Labs, was also a vice president of AT&T. There was only one person with superior authority, and that had to be the president, Walter Gifford. Consider the artistic posture of AT&T in 1929. Their late and revered former president, Theodore Vail, loved the arts and made it a company policy to support their development. Vail's protégé and handpicked executive, Gifford, became president in 1925 (Brooks 1975, 169). Gifford was also interested in art as evidenced by his role as founding trustee of the Grand Central Art Galleries. Although speculative, with this background in the arts it is likely that Gifford gave the orders to enlist outside artists, and there was no pushing back.

Thus artists were invited, and models were prepared by John Vassos, René Clarke, Lucian Bernhard, and Gustav Jensen. Related patents were issued and assigned to Bell Labs (Bernhard 1930; Clarke 1930; Jensen 1930; Vassos 1930). These artists worked independently and did not have close working relationships with Bell Labs engineers. In a 1932 report, Douglas King discussed each model and not surprisingly concluded that none was suitable for production (King 1932, 9). An example of these designs by John Vassos (fig. 14.1) exhibits superfluous decoration, as did most of the artists' designs, and shows the influence of the Art Deco period.

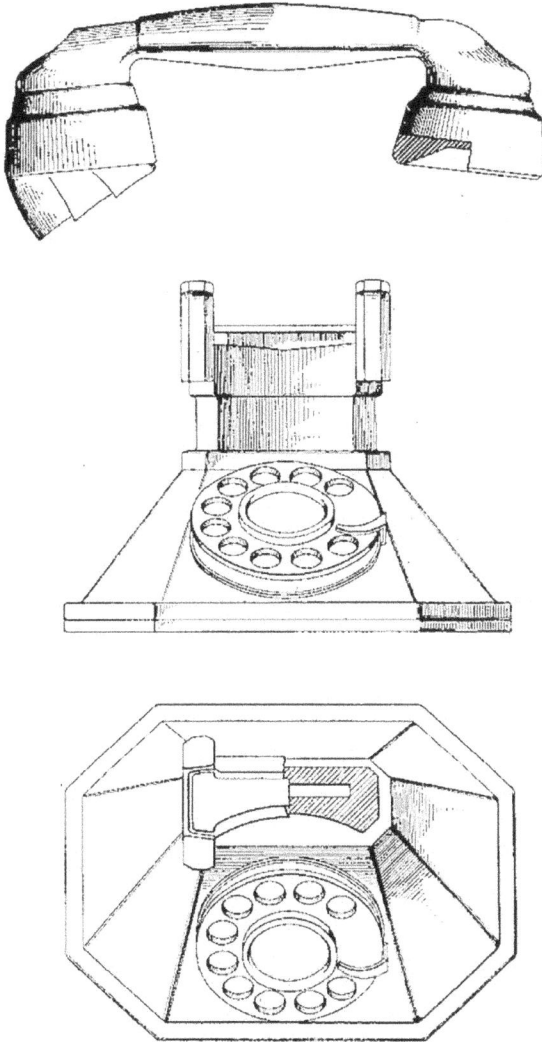

FIGURE 14.1 Selected drawings from Vassos's design patents.
(Reprinted from Vassos 1930.)

FIGURE 14.2 Photograph of Jensen's desk set model,
1930. (Reprinted from Wilson 1937, 135.)

Several other similar designs were patented in this time period by Bell System employees, who presumably had artistic interests (Blount 1930; King and Martin 1930; Labaugh 1930). These designs were all for desk and wall mountings that required separate subsets (as specified in Edwin Colpitts's 1928 letter) and therefore were not related to Bell Labs' next project of designing a "combined" (i.e., a complete) telephone—the Western Electric No. 302.

One of the outside artists' models, Jensen's, did not have a corresponding patent for the desk stand, and this omission might have been intentional. At about this same time, the Bell System adopted an anti-sidetone circuit designed by George Campbell, and they intentionally did not patent that circuit as had been done for the previous booster circuit. Instead, they published all of Campbell's circuits in such a way that a competitor could not in the future patent a variant that the Bell System might want to use (Fagen 1975, 107). Jensen's model (fig. 14.2), as it appeared in a 1937 article published in *Pencil Points*, was the only model that received some positive critique in King's summary report (King 1932, 5; Wilson 1937, 135). The postwar Princess telephone described in chapter 24 had many similarities to Jensen's model.

Dreyfuss later wrote that he had been informed that Bell Labs was offering $1,000 awards to each of 10 artists and craftsmen for their conceptions of the future appearance of the telephone (Dreyfuss 1955, 102). Notice that the 4 outside artists, the 3 inside designers, and Dreyfuss together brought the total to 8. It is reasonable to think that there were 2 other artists, who did not complete their work or left no record of it just as Dreyfuss had done, such that in all likelihood the offer was made to 10—in agreement with Dreyfuss's account.

Dreyfuss said that making such a design would require collaboration with Bell Labs technicians, but such collaboration was not offered because it was thought to limit a designer's artistic scope. Although Dreyfuss's further discussion on this offering suggests that he might not have accepted, he did not actually say that. In fact, Dreyfuss appears to have submitted a model to AT&T in 1930, but no picture of his model has been found. Our single reference comes from Dreyfuss's personal record of information on himself, the firm, and its clients, which was referred to as the "Brown Book" and is now in the Smithsonian Libraries collection at Cooper Hewitt Library (Dreyfuss 1969, 6). The entry shows the year as 1930, the client as "A.T. &T.," and the work done as "Model." We are not aware of any further factual information on this narrative.

Since working at Dreyfuss Associates in 1990–1991, author Flinchum always thought that Dreyfuss had shown a model to AT&T, found they didn't like it, took it back to the office, and had it destroyed before it could be photographed. Flinchum's basis was that he knew Dreyfuss had produced a model, but a thorough search of Dreyfuss's records did not produce any more evidence than that entry in the Brown Book. More recently, Jan Hadlaw reported a similar story that Dreyfuss had submitted a design (Hadlaw 2017, 144). However, in an earlier thesis on which that report was based, Hadlaw said, "I have not found any evidence in AT&T's files of correspondence with the consulting artists of an offer being made to Dreyfuss in 1929" (Hadlaw 2004, 79). We have not been able to find the basis for Hadlaw's story about a Dreyfuss model. It is possible that Hadlaw picked up this story—a conjecture—at Cooper Hewitt, where Flinchum curated a Dreyfuss exhibition in 1997 and Hadlaw later visited prior to 2004, as they interacted with the same people at the museum.

The Bell System's misstep was compounded by the fact that by 1929, when the artists' competition for a desk stand design was launched, telephone components had already been reduced in size such that a combined telephone could have been produced. The so-called combined telephone was self-contained with all components mounted in the desk set. Telephones had already begun to be produced worldwide in this combined configuration, and as with the conversion to dial phones, the Bell System was not the leader. It is argued that the Bell

System put off switching to dial telephones because they were concentrating on developing universal (wide-area) service, which is a reasonable defense. It also could be argued that the Bell System was not first with a combined telephone because the fiasco with the desk stand delayed going straight from the B-type base to a combined set.

15

THE BELL SYSTEM'S AMERICAN COMPETITORS FOLLOW SUIT

B y the end of the 19th century, the 17-year period of the original Bell patent had expired and competitors were on the scene legitimately in America. There were, by then, about 85 of these independent telephone manufacturing companies, including the Stromberg-Carlson Telephone Manufacturing Company (Stromberg-Carlson), the Kellogg Switchboard & Supply Company (Kellogg), and just a year later the Automatic Electric Company (Automatic Electric) (Pleasance 1989, 47, 111, 116, 124, 145, 201).

The Bell System is often portrayed as dragging its feet on introducing handset telephones, but they were in fact first with a modern molded handset—the Type A handset of 1924. The Bell System had paved the way with patents on strength and acoustic properties and made the head measurements that defined important dimensions. Because of their success, molded handsets would be present on all telephones for decades to come, and companies outside the Bell System jumped on the bandwagon with Bakelite designs.

FIGURE 15.1 Automatic Electric No. 1 Monophone desk stand, 1925. (Photo by Marketa Ebert.)

AUTOMATIC ELECTRIC NO. 1 MONOPHONE

Automatic Electric introduced its first handset desk set (fig. 15.1) at a convention of the Independent Telephone Association on October 21, 1925 (Automatic Electric 1925, 67). The desk stand's base was identical to that of their candlestick desk stand with a cradle mounted on top. Notice that the shaft on which the cradle was mounted was not tubular, however, and this shaped feature seemed to disguise the fact that the base was that of a candlestick phone. The dial was mounted on the surface of the base, and the desk set was in fact quite similar to the Bell System's B-type base that had been criticized.

Herbert Obergfell was Automatic Electric's longtime chief engineer and principal designer. We believe he was the designer of the original handset called the Monophone because his name appeared on earlier and later telephone patents for Automatic Electric. It is likely that Automatic Electric was unable to patent this handset because of the Bell System's previous patents on similar handsets, but Obergfell did obtain a patent on a later variation of the Monophone handset (Obergfell 1934a).

From table 11.1, critical dimensions of the Automatic Electric No. 1 handset can be seen to differ significantly from Western Electric's Type A handset. Figure

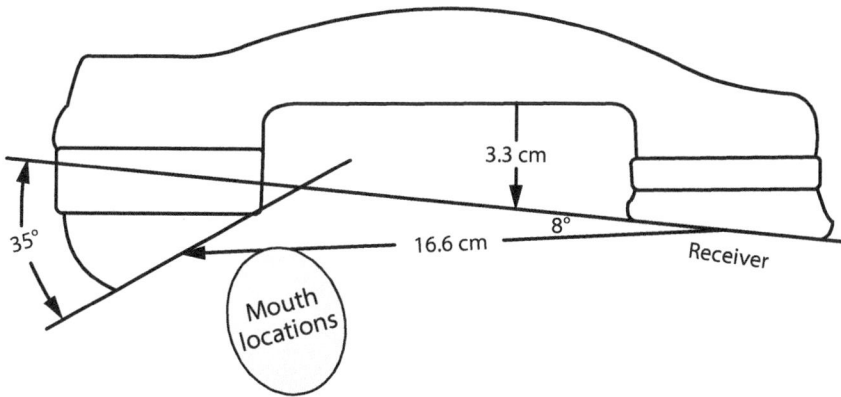

FIGURE 15.2 Outline and modal dimensions of the handset
for the Automatic Electric No. 1 desk stand.

15.2 shows the location of most mouths as measured by the Bell System. The Automatic Electric handset is thus seen to be long and the modal angle too shallow for the transmitter to be close to mouth locations. Although the Monophone handset was similar in appearance to the Western Electric Type A handset, it was not copied from that design, as can be seen from the dimensions.

Automatic Electric had switched to using the booster circuit shortly after the Bell System's patent expired in 1918 and continued to use it with the Monophone. As the Bell System had found out, a heavy molded Bakelite handle was sufficient to prevent acoustic feedback. But unlike the Bell System, Automatic Electric did not yet have a position-independent transmitter, so the Monophone telephones suffered from that shortcoming as well. Automatic Electric did not introduce a position-independent transmitter until 1935 (Telephony 1935).

STROMBERG-CARLSON NO. 1178

Stromberg-Carlson introduced its first molded Bakelite handset on a desk stand in 1929 (fig. 15.3). Ray Manson was chief engineer at Stromberg-Carlson at that time and was probably responsible for this design. The handset appeared to be visually similar to the Western Electric Type E handset, although its modal length was shorter (see table 11.1). The short length put the transmitter quite close to mouth locations, however, and this was favorable for performance. Stromberg-

FIGURE 15.3 Stromberg-Carlson No. 1178 desk stand, ca. 1929.
(Photo by Edward Maddox.)

Carlson had also switched to using the Bell System's booster circuit, although a position-independent transmitter had not yet been designed for this handset either.

KELLOGG NO. 730

In 1931 Kellogg introduced the first molded Bakelite handset in America that was streamlined. It no longer looked like a handle with bells hanging from the ends. This handset was both attractive and fit well to most heads by approximating dimensions from the Western Electric Type E handset (see table 11.1). The handset was called the Masterphone, and its design patent was shared by George Eaton and Royal Horn (Eaton and Horn 1931). Eaton was the chief engineer at Kellogg and Horn was the chief draftsman. The Masterphone was used on a No. 730 desk stand (fig. 15.4) with the base from their earlier candlestick desk stand. Initially the Masterphone was equipped with an old-style transmitter and used the booster circuit, but by 1933 this handset was backfitted with Eaton's important position-independent transmitter—and the Kellogg telephones also began using an anti-sidetone circuit at that time (Eaton 1935).

Thus we see that American handset desk sets of the 1920s and 1930s followed the Bell System format of mounting a molded cradle on top of a candlestick base

FIGURE 15.4 Kellogg No. 730 desk stand, ca. 1931. (Photo by Carroll Haugh.)

rather than following the European French phone style. Kellogg got out ahead with its streamlined handset that had critical dimensions similar to the data-based Bell System handset. Automatic Electric, on the other hand, appeared to have ignored the Bell System's work and produced a handset that could be criticized.

COLORS CONSIDERED

By early 1930, Bell Labs management had shown an interest in colored telephones, and the Western Electric B-type and D-type desk stands were made available in nine different colors in addition to black (Jones 1930, 1; Western Electric 1929, 101; 1935b, 101). The desk stands were made of metal, so painting was the obvious method of coloring. The Type E handset, however, was made of Bakelite, a phenolic plastic that was usually black or dark brown because of impurities in the starting materials. Colored Bell System handsets were thus also painted. The Western Electric No. 302 and the Type F handset were also available in nine painted colors plus black (Western Electric 1939, 117, 270).

At about the time that Automatic Electric introduced a new phone in 1934, both their Type 1A (not quite the same as the No. 1 described above) and the new Type 34 Monophones became available in nine colors in addition to black (Automatic Electric 1931, 10; 1934, 4). These two phones had desk sets and handsets made of Bakelite, and Automatic Electric took a different approach from painting. With dyes, they were able to make their mahogany, walnut, and black colors out of Bakelite. For the lighter and brighter colors, however, they switched from a phenolic plastic to a cellulose plastic that could be made quite clear and readily accept colored dyes. One brand of this plastic called Lumarith was introduced in 1926 by the Celanese Corporation of America. Another called Tenite was developed in 1929 by Eastman Chemical Company. Lumarith and Tenite were both used in telephone manufacturing.

There were unforeseen problems, however. While phenolic plastic is a stable synthetic material, cellulose is a biological product and it deteriorates with time. This led to shrinkage and distortion with age, and one formulation of Tenite used by the Bell System in the 1950s developed a strong odor. There were a lot of variables in the manufacture of cellulose plastics, however, and the odorous phone problem—along with the shrinkage problem—was solved by making changes in manufacturing.

Although colored phones were available quite early, it was recognized that providing them would multiply the number of articles of merchandise that must be cataloged and stocked, which would increase costs. Thus basic black remained the dominant color until well after World War II, when an almond color (somewhere between light beige and dark ivory) became popular.

PART 5

THE ART DECO AND BAUHAUS PERIODS

T he Art Deco period got underway following the 1925 Exposition Interna-
tionale des Arts Décoratifs et Industriels Modernes (International Exhibition of Modern Decorative and Industrial Arts) in Paris, and the Bell
System was recovering from its design brouhaha in 1929, when the American
stock market crashed. There was an initial increase in the Bell System's business
as frantic calls were made, actually boosting its revenue, and at the end of 1930
there were more than 15 million telephones in service in the Bell System, then
an all-time high (Brooks 1975, 187–99).

However, as hard times prevailed during the Great Depression, customers
started canceling service and the Bell System began to lose revenue. By the end of
1932, more than 10 percent of its customers had been lost and the Bell System had
to do something. Manufacturing took the hit. Western Electric laid off almost
80 percent of its workforce. Bell Labs, however, merely did belt tightening. They
suspended hiring, reduced salaries, and shortened the work week to four days
(Gertner 2012, 36). Bell Labs staff had recently won a Nobel Prize and their rep-
utation was at stake, so within a year their funding was restored. Thus the Great
Depression had little effect on innovation and design that took place at Bell Labs.

16

THE ICONIC PREWAR BELL SYSTEM TELEPHONE

B y 1930, complete desk telephones were showing up in Europe, and Bell Labs began work on its own design. George Lum would be its principal designer. George Renwick Lum (1882–1982), was born in Stamford, Connecticut, and attended the Julliard School of Music (*Colorado Springs Gazette Telegraph* 1982). Before joining Western Electric in 1918, Lum spent six years with Yale & Towne Manufacturing Company on mechanical and ornamental iron and bronze design, a year and a half with Duffner & Kimberly Company on leaded glass windows and lighting fixtures, and then six years in business for himself in similar work. During his early years with Bell Labs, he took courses at City College of New York in design of electrical equipment, spent two years at the Art Students League, another two years of study with Mahonri Young on sculpture and drawing, and several years with Michael Jacobs, drawing and painting (Bell Labs 1943a, xxiii). Lum, a Bell Labs employee, thus had plenty of training in the arts and was capable of telephone design.

The complete desk telephone was called a combined telephone because it combined all the components in the base of the desk set and did not require a separate subset. In early 1930, a senior department head at Bell Labs, R. L. Jones, asked the laboratories' vice president, H. P. Charlesworth, for approval to proceed with a specific program for the next telephone (Jones 1930). In that memorandum, Jones recommended (1) assigning two or more Bell Labs employees who had artistic training to study form, color, and appearance of the future telephone; (2) retaining the services of an outside consultant; and (3) setting up an advisory committee of active professional designers. This committee of three (Harvey Wiley Corbett, Ralph Walker, and Lee Simonson) was referred to as the Advisory Art Committee and should not be confused with the four outside artists (Vassos, Clarke, Jensen, and Bernhard) mentioned in chapter 14. Jones further stated that he would not proceed to contact the people they had in mind for this work until this program was approved.

Thus we know that work had not commenced on the combined telephone as of the date of this memo, February 18, 1930. Although Jones in his memo did not name the people to be on the design team, we know they were George Lum and Nelson Blount, both Bell Labs employees, and Henry Dreyfuss, an outside consultant. Lum received all of the design patents related to the No. 302 telephone as well as the regular patent for details of the desk set. The only relevant patent he didn't receive was the one that Blount received for details of the handset.

Blount was also a Bell Labs designer and had received several patents in 1930 for Art Deco designs related to AT&T's artist competition (Bell Labs 1943b). It was common for prewar industrial designers to have studied art and design in Europe, and Blount is known to have visited Europe at least two times as a young man. Then during the period 1930–1936, records show that Blount made numerous trips to the Hawthorne manufacturing facility with W. C. Jones in connection with the development of transmitters and receivers (Bell Labs 1929–1932). And in 1936, Blount filed a patent for the detailed design of the Type F handset for the No. 302 telephone (Blount 1938).

The consultant, Henry Dreyfuss (1904–1972), was born in New York City, attended a high school that was rich in artistic studies, and studied privately with the well-known stage designer Norman Bel Geddes (Flinchum and Meyer 2022, 4, 6). Young Dreyfuss, like Blount, had traveled to Europe in connection with his education. In 1930 Dreyfuss married the daughter of a former Manhattan borough president, and she instantly became his business partner. With his well-connected wife, they propelled their fledgling industrial design business into world fame in less than a decade, but this could not have been anticipated when Bell Labs hired him as a token consultant.

Dreyfuss appeared on the Bell System's radar in 1930 when, we believe, he produced a model during the artists' competition for a telephone design. He was thus recognized as an artist by AT&T, and his apparent refusal to carry through likely endeared him to Bell Labs. He thus would have been a suitable candidate for Bell

Labs to hire in deference to AT&T's desire for artistic attention. On August 27, 1930, Dreyfuss's secretary wrote to Blount to inform him of Dreyfuss's schedule for consulting meetings at Bell Labs during September (Hart 1930). Thus we know that work had commenced on the combined telephone by August 27, 1930, since the people involved had then been contacted.

But in 1930, 26-year-old Dreyfuss was just gaining recognition as a Broadway stage set designer. Dreyfuss, Bel Geddes, Raymond Loewy, and Walter Dorwin Teague would later be recognized as pioneers of industrial design; however, all had identified themselves as industrial designers only within the previous three years. Industrial design was not well developed in 1930 when Dreyfuss started consulting with Bell Labs, so it would not be correct to say that Dreyfuss brought the discipline to the Bell System. Several facts are relevant. First, Dreyfuss was a very talented designer. Second, Bell Labs had overlooked the value of aesthetics in their design of the B-type desk stand. Third, AT&T executives had overestimated the value of designs that artists could produce when working in isolation. A better approach would have been to find in-house engineers with artistic talent, as had been done the previous year, or to hire outside artistic talent to work closely with the engineers. This was a lesson that had been learned by Bell Labs before Dreyfuss was hired as a consultant. It is fair to say that Dreyfuss adopted this lesson from Bell System experience, and we know that he embraced it as his mantra: "[An industrial designer] consults closely with the manufacturer, the manufacturer's engineers, production men, and sales staff ..." (Dreyfuss 1955, 24). Dreyfuss swiftly converted his skill as a successful stage set designer to that of an industrial designer in the stimulating environment of the Bell System, and working together the discipline matured.

By August of 1930, Bell Labs would have been aware of the Siemens & Halske Model 26 (see chapter 17) and the Fuld & Company Frankfurt phone (see chapter 27), both of which were introduced in 1928. The Siemens Brothers Neophone (see chapter 17) had also just been introduced in 1929. Herbert Shreeve, who had worked for Edwin Colpitts when they were at Western Electric, brought back from Britain five Neophone handsets with cradles (i.e., complete desk telephones), as well as papers, drawings, et cetera, and they were sent to Bell Labs for examination (Colpitts 1929b). Ericsson's 1931 model was announced in the *Ericsson Review* at the end of that year (L.M. Ericsson 1931, 266). It is likely that Bell Labs got their first glimpse of the Ericsson design when this *Ericsson Review* arrived (by boat) in New York in late January or early February 1932. (Possible influence of these earlier designs on the No. 302 is discussed in chapter 18.)

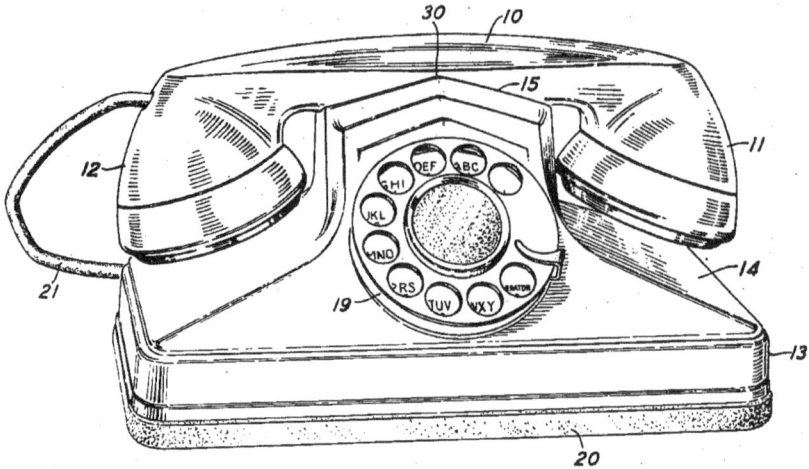

FIGURE 16.1 Drawing from Lum's patent. (Reprinted from Lum 1934.)

In May 1932, Lum filed a patent application for a combined desk set as shown in figure 16.1 (Lum 1934). Although we are not aware of any prototype of this design, it appears to have been the basis for the later Uniphone of Western Electric's Northern Electric manufacturing partner in Canada (fig. 16.2). The Uniphone was introduced to independent telephone companies in 1935, but Bell Canada didn't want it, stating that "a divergence from Bell System of standards ... we are very loath to consider at this time" (Bell Canada n.d.). Notice that the Uniphone is basically an integrated cradle on top of a small subset, which is oriented with its long dimension from side to side.

Later in 1932, Lum filed an application for a design patent and a detailed patent on a desk set as shown in figure 16.3 (Lum 1932, 1935c). This design was produced as a prototype (fig. 16.4), and the specimen shown is dated IV32, the fourth quarter of 1932 (Fassbender 2011, 8). The design of this prototype differs from Lum's other design in two significant ways. First, the handset cradle is integral with four prongs. Second, the orientation has changed from the long dimension being from side to side to this dimension being from front to back, thus reducing the usable desk space that would be required. Both of these features would carry over to the production No. 302.

Three years later, on March 27, 1935, Lum filed applications for a design patent and a regular patent on the No. 302 desk set (Lum 1935a, 1938). One of his design patent drawings is shown in figure 16.5, and a production No. 302 telephone

FIGURE 16.2 Northern Electric's Uniphone telephone, 1935, with an early handset. (Photo by Matthew Gay, courtesy of the Gregg Museum of Art & Design, North Carolina State University.)

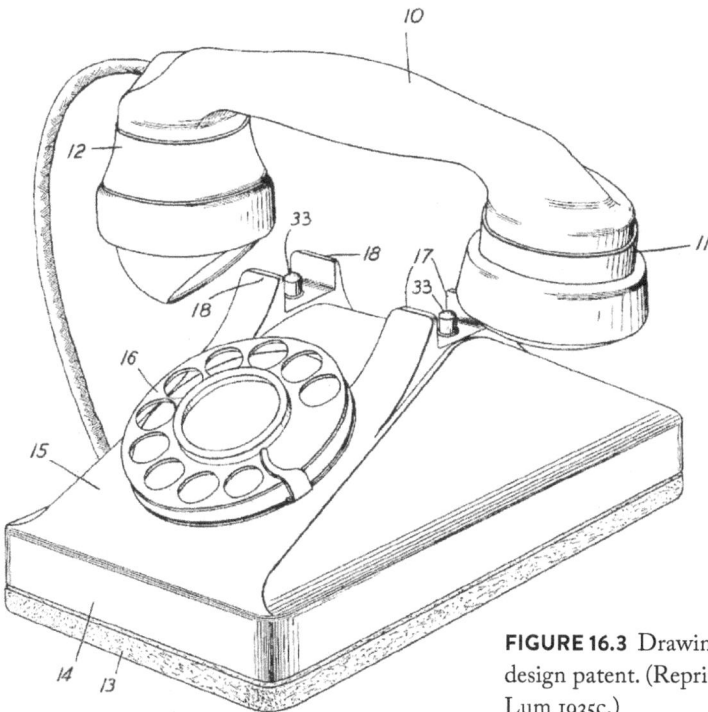

FIGURE 16.3 Drawing from Lum's design patent. (Reprinted from Lum 1935c.)

FIGURE 16.4 Western Electric prototype desk telephone, 1932.
(Photo by Paul Fassbender.)

is shown in figure 16.6. Lum's design patent application for the Type F handset was filed soon thereafter, on April 25, 1935, but he did not simultaneously file an application for a detailed patent for the handset as he had done for previous designs (Lum 1935b). Instead, Blount filed for the detailed patent, and his application for the regular detailed patent was not filed until September 26, 1936 (Blount 1938). These dates suggest that there were some last-minute details on the handset that were not resolved until late 1936, and this is consistent with the 1937 introduction date of the No. 302 telephone.

Although produced during the Art Deco period, the No. 302 displayed no added decorations that would be characteristic of that genre. Instead, its design was more functional in line with the Bauhaus school. There were, nevertheless, several design features of the No. 302 that deserve special mention, and they are related to the handset and a lifting handle.

A comparison of the Type F handset for the No. 302 telephone and the earlier Type E shows that the modal distance and angle—and even the tilt of the transmitter—were almost the same for the two handsets (fig. 16.7). Thus the new handset did not deviate from the earlier anthropometric measurements. The extensions of the transmitter and receiver from the handle were a little shorter in the later handset because the transmitter unit and the receiver unit were then smaller, yet clearance was still greater than the distance-to-cheek measurements. By the way, the Type F handset was intended to be interchangeable with the Type E handset on the No. 302 and earlier desk stands, so this similarity in dimensions was intentional (Jones 1938, 341).

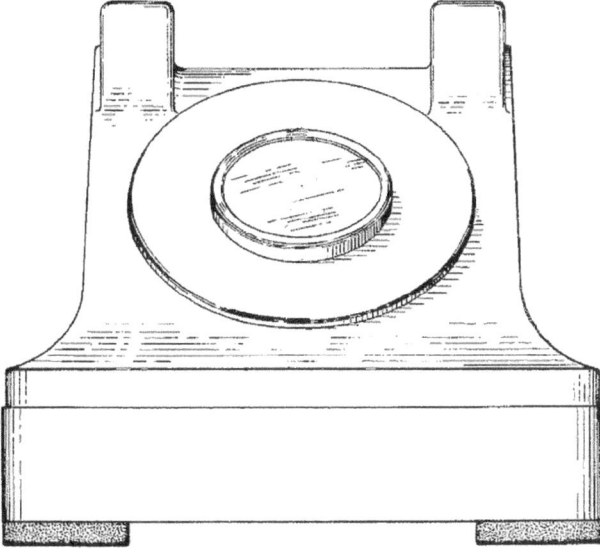

FIGURE 16.5 Drawing from Lum's design patent. (Reprinted from Lum 1935a.)

FIGURE 16.6 Western Electric No. 302 desk telephone, 1937.
(Photo by Sally Andersen-Bruce, courtesy of the US Postal Service.)

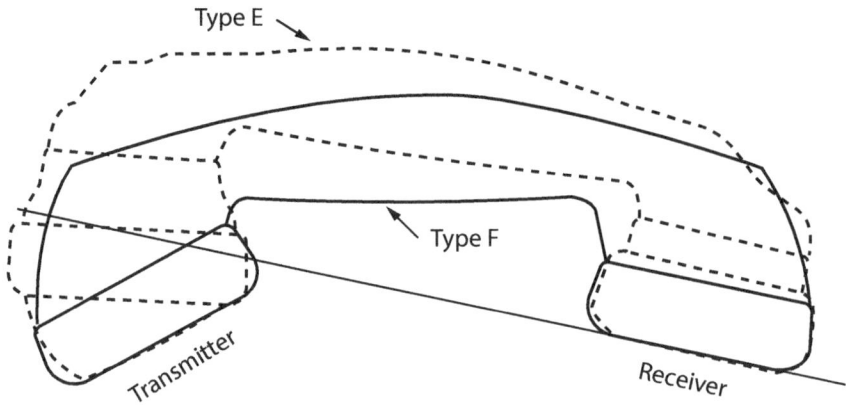

FIGURE 16.7 Comparison of the Western Electric Type E and Type F handsets.

In Lum's handset design of 1935, there was a feature that we credit to Dreyfuss, who was acting as a consultant to Lum at that time. It was small, but interesting. The ridge on top of the handle (fig. 16.8) came to a prominent peak at both ends, and this feature was totally absent from Lum's earlier 1932 design as well as the Type E handset with similar triangular central cross sections.

Looked at in context with other work Dreyfuss was doing during this period, a center ridge appears to be a theme of his work in the mid-1930s. Three other designs of that period from the Dreyfuss firm had this feature and are shown in figure 16.9 (left to right): Minneapolis Honeywell Acratherm thermostat, 1935; John Deere tractor, 1938; and 20th Century Limited train, 1938 (Dreyfuss 1935, 1938, 1939).

While Dreyfuss himself probably designed the thermostat, we believe that Herbert Barnhart of the Dreyfuss firm designed the tractor and the locomotive, and we are sure that Lum of Bell Labs designed the telephone; nevertheless, Dreyfuss was a common denominator and likely suggested this motif to each designer (Flinchum and Meyer 2022, 40, 81, 87). Also note that these four designs went into production within a very short period of time, 1937–1938. As with many of his competitors of the 1930s, Dreyfuss was a streamliner, but avoided more radical possibilities in favor of visual durability—which also characterized his later work.

A subtle, but very important, additional design feature of the No. 302 telephone was the cavity behind the cradle that formed a perfectly balanced lifting

FIGURE 16.8 Pronounced ridge on the Western Electric Type F handset of 1935. (Photo by Ted Mueller.)

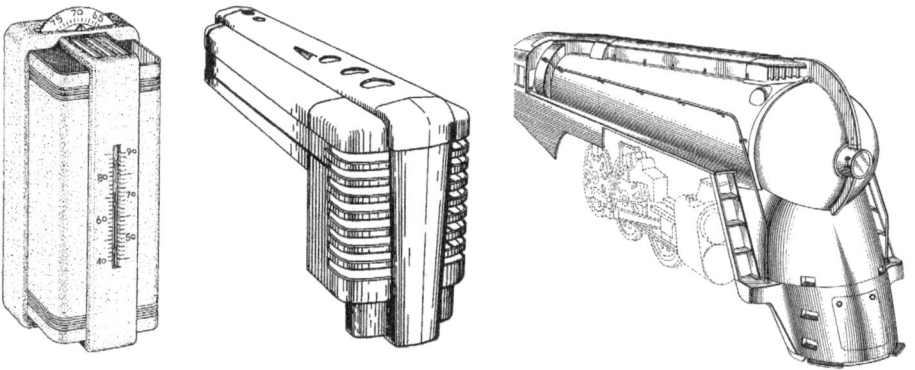

FIGURE 16.9 Dreyfuss's patent drawings for mid-1930s designs that had a central ridge. (Reprinted from Dreyfuss 1935, 1938, and 1939.)

handle (figs. 16.10 and 16.11). A lifting handle of some form was built into every one of Dreyfuss's later desk telephones, so we also credit this feature to Dreyfuss, the consultant.

Not long before the design of the No. 302 was finalized, Frank Jewett, vice president of AT&T, wrote to Edgar Bloom, then president of Western Electric, about the lessons learned from the difficult history of replacing the unattractive B-type desk set (referred to as AT&T Case No. 34648). Among other things, Jewett said the following (Jewett 1935, 2):

> I have advised Mr. Colpitts and others at the Laboratories that in the future I think the Laboratories should exercise its responsibility for design in a more positive fashion and should decline to undertake redesigns, involving delay and expense, which do not affect the proper functioning of equipment and which affect appearance and convenience to a minor extent only.... [We] have established the procedure of making the Laboratories designers responsible for the actual housing designs. As an aid to those in their work we have likewise established a small advisory committee of leading industrial artists and architects who from time to time are asked

FIGURE 16.10 Lum's patent drawing shows the cavity behind the cradle that forms the lifting handle. (Reprinted from Lum 1935a.)

FIGURE 16.11 Western Electric No. 302 telephone being lifted by its carrying feature.

to meet and comment on the Laboratories designs. They are not asked to assume any responsibility but merely to give us the benefit of their expert judgment on what we are proposing to do. So far this scheme has worked very satisfactorily and for the present at least I intend to continue it.

Bell Labs thus grabbed control over future designs, acknowledging that they had been jerked around by corporate headquarters in responding to appearance complaints. Bell Labs continued to hire Dreyfuss as a consultant, but we found no further information on the Advisory Art Committee. Then after the war—and before any future designs could be made—AT&T took this responsibility away from Bell Labs and gave it to their design consultant, Henry Dreyfuss, with whose firm it would remain for the rest of the Bell System's lifetime.

17

NEW INTERNATIONAL
DESIGNS EMERGE

S everal European and American telephones in addition to the No. 302 were introduced during the Art Deco and Bauhaus periods. Some of the more prominent ones are described below.

SIEMENS & HALSKE MODEL 26

The 1925 exposition in Paris had political overtones that restricted invitations to allies of France during World War I. Obviously, Germany was not invited. This probably contributed to Germany's anti–Art Deco sentiment, because its styles were more aligned with the Bauhaus movement. Walter Gropius led that program in Germany in the Bauhaus school founded in Weimar in 1919. The modernist principle championed by the Bauhaus movement was that form follows function. Such designs did not display the unnecessary decoration of the Art Deco style.

Similar to the Bell System in America, the German Post Office (Reichspost) owned telephones and provided service to subscribers. When the relatively modern-looking Automatic Electric No. 1 desk set came out in 1925, the official phone in Germany still looked like the old square Siemens phone shown in figure 9.5. The Reichspost went out for bids for a new telephone and in 1928 introduced

FIGURE 17.1 Siemens & Halske Model 26 telephone, 1926.
(Photo by Arwin Schaddelee.)

a design by Siemens & Halske (Museumsstiftung 2001, 123). Siemens & Halske had designed the telephone in 1926, and it was the ancestor of a family of telephones that would remain in service in Germany for almost four decades (fig. 17.1).

The Model 26 was the first of the emerging desk telephones that contained all of the working parts in a single integrated case, although there had been at least one complete telephone with a desk set cradle mounted on an existing subset box (Obergfell 1925). While this German phone was created during the Art Deco period, it was definitely not of the Art Deco style, though no formal connection has been found with the Bauhaus school.

The Model 26 telephone was remarkably modern looking for its time. We are not sure whom to give credit for this design, although probably Otto Weeber and Otto Soldan. A substantial number of German, Austrian, and American patents were issued for the Model 26 telephone, but the German and Austrian patents were awarded to companies rather than individuals and did not identify who the designers were. However, a US patent was given to Weeber for the handset of this phone, and Weeber and Soldan together were given US patents for a related telephone (Weeber 1928; Weeber and Soldan 1931, 1934). According to Claudia Salchow of the Siemens Historical Institute, no designers were

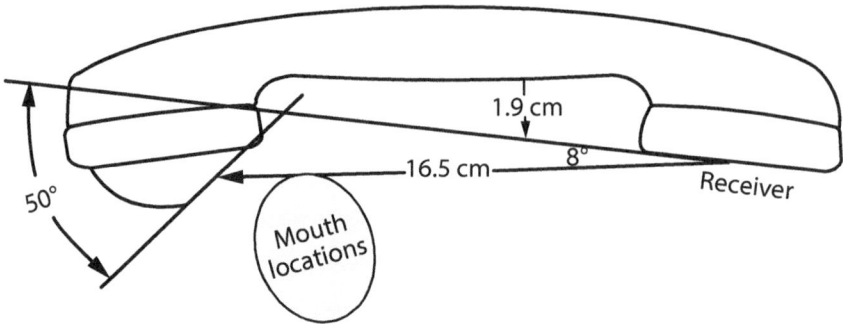

FIGURE 17.2 Outline and modal dimensions of the handset
for the Siemens & Halske Model 26 telephone.

employed at Siemens at that time, so it is likely that Weeber and Soldan were
engineers with artistic talents, as we have found to exist in other design efforts.

The handset of the Model 26 (fig. 17.2) is revealing. By 1925 there were only
two other molded Bakelite handsets in existence: the Western Electric Type A
of 1923 and the Automatic Electric No. 1 Monophone of 1925. Neither of those
designs had a streamlined appearance such as the Model 26, and another streamlined handset would not come along until the Kellogg Masterphone of 1930 (see
chapter 15 and below). Nevertheless, the handset on the Model 26 does have some
shortcomings. For example, the handle is long and the angles are shallow, placing the transmitter far from a user's mouth, and the handset is too slim to provide adequate finger clearance for users with ample cheeks.

Like the Automatic Electric No. 1 Monophone, the Model 26 design was
completed before the Bell System published in 1932 the results of its earlier anthropometric study. The Western Electric Type A may not have been readily
available for examination at Siemens & Halske because it was a prototype and
was not offered for sale by the Bell System. But Siemens & Halske should have
known by 1925 that the Bell system had conducted that study—4,000 subjects
whose heads were measured could not have kept a secret for five years—and existing patents for the Type A handset might have been examined to obtain its
important dimensions. Instead of copying Western Electric's dimensions, Siemens & Halske appears to have copied the Automatic Electric No. 1 dimensions (see table 11.1).

Further, the transmitter in the Model 26 appears to be smaller than the Western Electric transmitter—another shortcoming. Based on our measurements,

the unclamped area of the diaphragm in the Model 26 transmitter was about 25 percent smaller than that in the Western Electric Type E handset. Because the forces acting on the carbon granules in a transmitter are proportional to the diaphragm's area, the transmitter was thus about 25 percent less efficient than the Western Electric transmitter. Siemens & Halske continued using these dimensions, including the smaller transmitter, in later variations of the Model 26 design, although the finger clearance was increased a little.

SIEMENS BROTHERS NEOPHONE

Shortly after Werner Siemens and Johann Halske formed their telegraph works in Berlin, a branch was opened by William Siemens near London. Halske later withdrew from business in the London branch and it was renamed Siemens Brothers Telegraph Works (Siemens Brothers). In 1929, they produced a desk stand in Great Britain that could be converted to a combined phone by adding a base containing the extra components (Aldridge et al. 1929; Siemens Brothers 1931). This phone, called the Neophone by Siemens Brothers, is shown in figure 17.3 with the base attached and was adopted by the British Post Office to replace their candlestick telephones.

FIGURE 17.3 Siemens Brothers Neophone telephone, 1929. (Photo courtesy of BT Group Archives.)

The stepped pyramidal shape of the phone strongly suggests that its designer had some artistic experience, but the designer is unknown according to Andrew Emmerson, a prominent British telephone historian. Robertson's book on telephone history relates the following story about the design (Robertson 1948, 226):

> It is reliably recorded in the Siemens works, where the now standard type was evolved, that a leading engineer of the company—after a long period of ineffective designing—saw, one day, in a shop window in London an Edwardian silver ink stand (of the type, it is suspected, that Lord Curzon had, "made of crystal and gold," on his desk in his viceregal palace); its simple elegant curves pleased him; he bought it, and arranged its adaptation to telephone design. And the result all now know. The Post Office accepted it, and standardized it, but rejected Siemens' trade name for it, "the Neophone," and called it austerely the "Hand Combination Set."

Like Siemens & Halske, Siemens Brothers should have known by 1928 that the Bell System had conducted a major anthropometric study; they could have examined specimens of the Western Electric Type E handset and approximated its important dimensions. Instead of copying dimensions, Siemens Brothers could have at least inferred the location of typical mouths from the Bell System design. Nevertheless they made fresh measurements of a large number of heads and based the length of the handset on those measurements (Aldridge et al. 1929, 187). Figure 17.4 shows that Siemens Brothers did not copy the Bell System

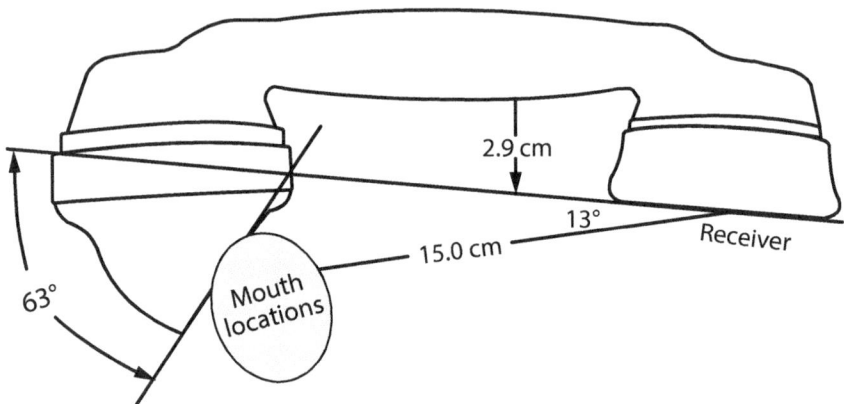

FIGURE 17.4 Outline and modal dimensions of the handset for the Siemens Brothers Neophone telephone.

dimensions (see table 11.1), but the location of mouths relative to the transmitter from the Bell System study was almost exactly the same in the Neophone handset as it was in the Bell System's Type E handset.

ERICSSON 1931 MODEL DE 702

Ericsson in Sweden was also developing a true combined telephone in 1929 (Brunnström 2006, 176). Design of this phone originated with Johan Bjerknes, who was the design manager for Ericsson's subsidiary in Norway, the Elektrisk Bureau. Bjerknes's initial concepts were given to the well-known Norwegian artist Jean Heiberg in September 1930. Thus Ericsson placed even more emphasis on aesthetics with this phone. By January 1931, Heiberg had produced a model in plaster, and Ericsson began production of this phone (fig. 17.5) in October 1931 (Grönwall 1933; L.M. Ericsson 1931). This rotary dial desk telephone was identified as model DE 702 but was more commonly known simply as the Ericsson 1931 model. By the way, a modified version of this phone was adopted in 1937 by the British Post Office as their then standard telephone (Freshwater 2022, 4).

FIGURE 17.5 Ericsson DE 702 desk telephone, 1931. (Photo by Matthew Gay, courtesy of the Gregg Museum of Art & Design, North Carolina State University.)

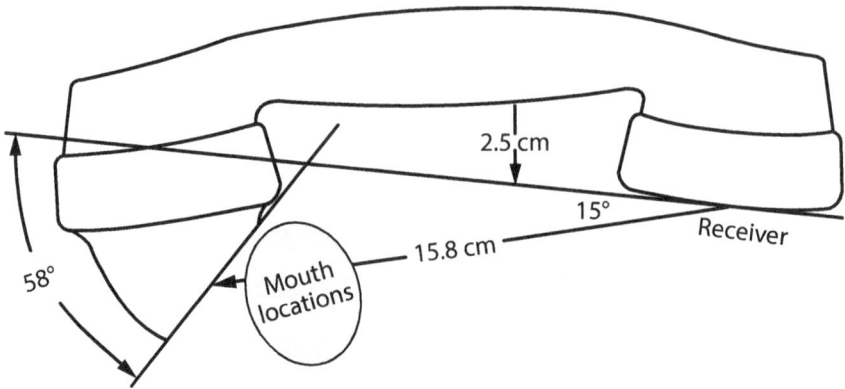

FIGURE 17.6 Outline and modal dimensions of the handset for the 1931 Ericsson telephone.

For this design, Ericsson appears to have been paying attention to both the Bell System and the Siemens Brothers designs. Ericsson produced a handset that was well positioned with regard to mouth locations (fig. 17.6), although the angles and dimensions were not the same as the others.

Nevertheless, the Ericsson phone had one significant weakness from a user's point of view. Telephones of that vintage were heavy—about 5 pounds (2.3 kg)—and the housings were invariably cast with a foreword draft (narrower at the top than the bottom) so they could be removed from the casting mold. Because of this shape, they were surprisingly difficult to pick up since they would slip through your fingers. This deficiency would be remedied in later telephone designs with a feature to permit grasping the phone.

THE BTMC NO. 2724 ANTWERP TELEPHONE

The American Bell System also had an early presence in Europe. In the same year Western Electric started making telephones for the Bell Company, 1882, the two companies opened a joint operation in Europe to manufacture and market telephones (Estreich and Verhelst n.d.). It was based in Antwerp, Belgium, and called Bell Telephone Manufacturing Company (BTMC). Their reach was intended to cover Europe but eventually extended to Asia as well. Thus the Bell System was on its way to world dominance, but it wasn't to be. Antitrust concerns emerged in America, and in 1925, BTMC was sold to International Telephone & Telegraph (ITT). Although the BTMC name was retained, by the beginning of the Art Deco and Bauhaus periods, all of the BTMC telephones were independent of the American Bell System.

A few years after the Siemens & Halske Model 26 was introduced, BTMC designed a very similar telephone, the No. 2724, whose housing was molded Bakelite (Pocock and Schreiber 1936; Schreiber 1935). This desk phone, commonly called the Antwerp telephone (fig. 17.7), had a companion wall phone that used the same base plate with components, which facilitated both manufacturing and servicing. This and other BTMC phones were marketed widely in Europe, Asia, South America, and Australia.

By this time, the Bell System had already patented a precursor to the next-generation Type F handset (Lum 1934). Although independent of the Bell System, the handset on the Antwerp telephone is similar in appearance to that of the Bell System's phone, but again a little longer (see table 11.1). It is odd that most European designs continued to have longer handsets than their American counterparts because the shorter handsets placed the transmitter closer to the mouth and produced stronger transmitter signals. Perhaps articulation was better with the transmitter a little farther from the mouth, and strong transmitter signals were not needed for the shorter lines in Europe.

FIGURE 17.7 BTMC No. 2724 Antwerp telephone, 1935.
(Photo by Remco Enthoven.)

FIGURE 17.8 Kellogg No. 900 Masterphone telephone, 1936.
(Photo by Carroll Haugh.)

KELLOGG NO. 900 MASTERPHONE

In America, the Bell System's competitors also produced some interesting designs during this period. Perhaps the most unusual was the Art Deco–style Kellogg No. 900 non-dial desk set of 1936 (fig. 17.8). This set was designed by Henry Billington, a sales manager at Kellogg, and it used the earlier Masterphone handset (Billington 1937). George Eaton's position-independent transmitter had been developed in 1933 and was incorporated in this handset for the Kellogg No. 900 (Eaton 1935).

We know little about the design process for this phone, but two holes were provided in the webbing beneath the handset cradle. By placing two fingers through those holes, the phone could easily be lifted with one hand with the handset in its cradle. The Kellogg 900 design was unable to accommodate a dial, though, and the design did not develop further.

AUTOMATIC ELECTRIC TYPE 34 AND TYPE 35

In 1934 and 1935, Automatic Electric introduced a pair of telephones that became quite popular. One was a desk phone and the other was a wall phone. Obergfell, the company's chief engineer, was awarded the design patents for both of these phones, and he also made changes to the earlier Monophone handset (Herbert

FIGURE 17.9 Shirley Temple using the Automatic Electric Type 34 telephone in 1938. (Photo courtesy of Rita Dubas.)

Obergfell 1932, 1934a, 1934b). The handset was shortened and the camber angle increased, bringing the transmitter much closer to typical mouth locations. The Bell System had just published its report on the company's anthropometric head measurements, but the timing was such that Obergfell must have copied major dimensions of the Western Electric Type E handset (see table 11.1).

The desk phone was Automatic Electric's Type 34 that was featured in a 1938 promotional article showing child actress Shirley Temple (fig. 17.9) with the 100,000th phone to be installed by her local telephone company (Dubas 2006, 115; Telephony 1938). Figure 17.9 is the keybook photograph of Temple, used by the studio for ordering prints and negatives. Temple was 10 years old when this picture was taken and missing a tooth. On the photo, the word "Retouch" can be faintly seen with a line and a half-arrowhead pointing to the missing lower tooth. The retouched version was used in the article.

Although the Telephony article was about a local telephone company's service growth, the Type 34 telephone became known as the Shirley Temple phone as a result of this publicity that spread nationally. The Type 34 desk phone had no provision for lifting and was, therefore, just as inconvenient to move around as the Ericsson phone. An updated version of this phone, the Type 40, had an optional handle that could be attached to the switch hardware. It was clever and effective, but it was an afterthought.

The companion Type 35 wall phone with striking Art Deco features is shown in figure 17.10. This phone also had a nickname: the jukebox telephone. An interesting sidelight of this telephone is that Henry Dreyfuss, who worked exclusively on Bell System telephones, used the Type 35 telephone in his interior design of the New York Central's 20th Century Limited luxury train—a train that made Dreyfuss famous. Although the exterior of this train is iconic (see fig. 16.9), it was the revolutionary design of its interior that was really special. One of the luxury interior features was a telephone in each compartment that could be used to call for service and make dining reservations. This phone needed to be fixed to a wall, and the Bell System had not developed a modern handset wall phone by 1938 when the train went into service. Thus, Dreyfuss had to use the phone of his client's competitor.

FIGURE 17.10 Automatic Electric Type 35 wall telephone, 1935. (Photo by Marketa Ebert.)

STROMBERG-CARLSON'S ART DECO ENTRY

Ray Manson, chief engineer at Stromberg-Carlson, drafted a bulletin dated September 10, 1934, entitled "Combining Beauty with Utility in Hand Set Telephones" (Manson 1934). He wrote extensively about the need to make phones pretty so people would quit hiding them and said in his draft that the design of the one-piece telephone was put up to the internationally known designer Everett Worthington of Chicago. Worthington's family said that he hand carved a wood model at his home and that wood was his favorite medium. Unfortunately, the wood model Worthington submitted (fig. 17.11) was not for a one-piece telephone but rather for a desk stand that required a separate subset. Keep in mind that Stromberg-Carlson was located in Rochester, New York, such that Worthington was not embedded in the manufacturing group, and this undoubtedly contributed to the design blunder.

Nevertheless, in the late 1930s production models (Nos. 1198 and 1212) were based on this design (Stromberg-Carlson 1940, 5, 8). One was a similar desk stand, which required a separate subset, and the other was a puffed-up rendition that was a complete desk telephone. In both cases almost all of the detailing in Worthington's design was eliminated from their bases that echoed the details on

FIGURE 17.11 Model of the Art Deco desk stand by Everett Worthington for Stromberg-Carlson. (Photo courtesy of the Hagley Museum and Library.)

the handsets. The only remaining Art Deco vestige was the faceted rim on greatly simplified caps for the transmitter and receiver. Important dimensions of the revised handset are essentially the same as the earlier No. 1178 (see table 11.1), so it was well positioned relative to the location of most mouths.

KELLOGG NO. 1000 MASTERPHONE

In 1946, Kellogg introduced a desk telephone that was the epitome of Art Deco streamlining (fig. 17.12). Four small grooves around its rubber feet, like the marquees at Radio City and the Apollo Theater, accentuate the Art Deco influence. The handset on this telephone, still called a Masterphone, was a stretched-out modification of the earlier Masterphone, and the added length would have pulled the mouthpiece away from the location of most mouths (see table 11.1). It appears that the increased length was needed to fit the larger body of the phone. The design patent for the body was applied for in 1946 by Clifford Erickson of Chicago, who doesn't seem to have attended college but studied architectural drawing in high school (Erickson 1948). Erickson probably went to work as a draftsman for Kellogg in Chicago straight out of high school.

The phone's early introduction after World War II suggests that the design was done before the war and that its introduction was delayed. Two months

FIGURE 17.12 Kellogg No. 1000 Masterphone telephone, 1946. (Photo by Carroll Haugh.)

before the war ended in Europe, Kellogg previewed the phone and said it was waiting to go on production lines as soon as conditions permitted (Telephony 1945). A handhold cavity was built into this design, but it was so far to the rear that the phone was not balanced when lifting. The delayed introduction probably shortened the lifetime of this beautiful phone because it was soon eclipsed by the postwar Western Electric No. 500 (see chapter 21). The unbalanced lifting feature, along with a propensity for cracking of the thin Bakelite housing, undoubtedly contributed to this design's short lifetime as well.

18

HENRY DREYFUSS
AND THE MYTHS

The design of the No. 302 telephone is usually attributed to Henry Dreyfuss and is said to have been inspired by the Ericsson desk telephone of 1931. We believe that both of these assertions are incorrect, and our basis is described below.

In his foreword to our book on the Dreyfuss firm, Roger Manley wrote that mistaken attribution "only underscores Dreyfuss's impact on our culture, for one of the sure signs of a recognized genius is the tendency of others to assign authorship to them of things that they never touched or thought of. A valid way to measure Rembrandt's fame is by the number of paintings attributed to him that he didn't actually paint, just as one can gauge the respect for Oscar Wilde's wit by all the clever quotes attributed to him that he never actually said" (Flinchum and Meyer 2022, xvii).

In Dreyfuss's case, he was widely believed to have designed the iconic Western Electric No. 302 that was introduced in 1937. This belief was even commemorated with a US Postal Service (USPS) postage stamp as recently as 2011 (fig. 18.1). We both made this mistake in the past, giving full credit to Dreyfuss (Flinchum 1997, 97; Meyer 2005, 77). The most prominent recent appearance of the error is by Ellen Lupton, who was senior curator of contemporary design at Cooper Hewitt, Smithsonian Design Museum (Lupton 2014, 22). If we try to trace the origin of the error to its source, we find that Lupton and Meyer cite Flinchum, Flinchum cites Hiesinger and Marcus, and Hiesinger and Marcus cite Wallance (Hiesinger

FIGURE 18.1 2011 USPS postage stamp celebrating pioneers of industrial design. (Courtesy of the US Postal Service.)

and Marcus 1993, 121, 326; Wallance 1956, 36). Ann Ferebee, in her well-known history of design, also makes this mistake but cites only AT&T Photo Service with no details (Ferebee 1970, 38).

Don Wallance was a prominent designer working at mid-century, and his book *Shaping America's Products* was published in 1956—just one year after Dreyfuss's *Designing for People* came out. After faithfully characterizing Dreyfuss's own vague description of his early work with Bell Labs, Wallance gratuitously added a sentence for which there was no similar information (or picture) in the Dreyfuss book: "The results of Dreyfuss' first collaboration with the Bell Laboratories is [*sic*] evident in the form of the Combined Handset of 1937 (fig. 16)." Wallance's fig. 16 was a picture of a Western Electric No. 302. He provided no basis for this sentence, which in fact had another flaw. The No. 302 telephone of 1937 was called a combined telephone rather than a combined handset. The No. 302 combined the components of a wall-mounted subset and a desk stand into a single instrument. These separate units were features of previous telephones.

And the Dreyfuss-like features—the handset ridge and the lifting handle—would not be evident, as claimed by Wallance, to anyone who has not studied Dreyfuss designs. Therefore Wallance's claim, that Dreyfuss's influence would be evident, is not correct. Dreyfuss himself became complicit in this myth when he published a new edition of *Designing for People* in 1967 and added a photo montage that appears to feature more of his designs without explicitly saying so (Dreyfuss 1967, 51). In the lower right-hand corner of that two-page montage is a picture of a No. 302. Two later editions (1974 and 2003) include the same montage (Dreyfuss 1974, 51; 2003, 273). Those of us who later inferred that Dreyfuss had designed the No. 302 should have checked primary references.

When we first discovered this error, we double-checked our conclusion by looking for consistency between the absence of claims on Dreyfuss's behalf and the presence of claims by the Bell System. We have found no mention of Dreyfuss in prewar Bell System publications. And a more recent statement is carefully worded to say that Dreyfuss had a hand in the designs, rather than he did the designs (AT&T 1967, 19). Dreyfuss was not shy about advertising his own work, however, and in fact he published records of his industrial designs from 1929 to 1967 (see reproductions of all five in the appendix of Flinchum and Meyer 2022). No mention is made in those publications of the No. 302, although the No. 500, the Princess, the Touch-Tone (the Bell System's trademark for the pushbutton dial) desk phone, and the Trimline are featured. (More on those designs below.) One of the authors of the present book (Flinchum) worked as a slide curator and archivist at Henry Dreyfuss Associates in New York for about a year (1991–1992) and found no records related to the 300 series of designs other than a few notes on appointments for early meetings that Dreyfuss attended at Bell Labs. This was in stark contrast to the numerous records related to the later telephone designs.

On the other hand, design patents for the No. 302 desk stand, the handset, and even a related wall phone were all awarded solely to George Lum (Lum 1935a, 1935b, 1949). We are aware that patents are sometimes granted to company executives or owners when the actual work was done by their employees. However, Bell Labs has been described as a meritocracy, and giving credit to others would have been contrary to the company's policy on patent assignments (Brooks 1975, 129; Brown 1954, 8). Further, Lum had a record of numerous other patents on ornamental designs, and on his 25th anniversary with the Bell System, the company explicitly gave him credit for designing the "handset and the combined set" (Bell Labs 1943a, xxiii). Additionally, it's clear that Lum's training in the fine arts was sufficient to enable such design work.

Another myth about the No. 302 telephone—that its design was inspired by the 1931 Ericsson phone—has also been perpetuated by Lupton, Flinchum, and Hiesinger and Marcus. The Ericsson phone, as shown in figure 17.5, has similarities to the No. 302, so inferring a connection is again understandable. But looked at from another angle (fig. 18.2), there is little resemblance to the No. 302 (fig. 18.3).

FIGURE 18.2 1931 Ericsson telephone, seen from the back. (Photo by Remco Enthoven.)

FIGURE 18.3 Western Electric No. 302 telephone, seen from the back.

By the way, the bracket on the back of the Ericsson phone is an accessory for parking a watchcase-style "mother-in-law" receiver. Further, inspiration by the Ericsson design would not necessarily be one's conclusion if other designs were available to suggest similar features, and there were three other designs in production at that time besides the Ericsson 1931 model.

Consider the Siemens & Halske Model 26 design (see fig. 17.1). Its handset is streamlined, in contrast to the older-looking handset on the Ericsson phone. The handset on the No. 302 is also streamlined—or clean-lined as Dreyfuss liked to say. Although this clean-lined feature might not have been influenced by the Model 26, it did not come from the Ericsson design.

Next, consider the Fuld & Co. Frankfurt model (see fig. 27.1) with a handset cradle that is fully integrated with the base. Although that phone's severe square design might be off-putting to some tastes, the cradle feature was revolutionary and appeared a full three years before the Ericsson phone. Thus the Frankfurt phone might have influenced this integrated feature on the No. 302. Though the Frankfurt model doesn't have four stationary prongs on its cradle, the ears on the No. 302 and its prototype do not look like those on the Ericsson phone either.

The Siemens Brothers Neophone (see fig. 17.3) had sweeping concave surfaces before they showed up on the Ericsson phone, and furthermore they were symmetrical on all four sides rather than just two sides as on the Ericsson phone. The Bell System's No. 302 has fully symmetrical concave surfaces, so it is likely that those shapes were influenced by the Neophone, which is known to have been studied by Bell Labs during the design of the No. 302.

Work on design of the housing of the Western Electric No. 302 was begun between February 18 and August 27, 1930. Initial concepts for the 1931 Ericsson phone were not given to the artist, Jean Heiberg, until September 1930 (Brunnström 2006, 176). Thus the Ericsson phone is unlikely to have inspired the Western Electric No. 302. Whether the 1931 Ericsson phone had any influence on the No. 302 is also problematic because the first glimpse of a photograph of the Ericsson phone came to Bell Labs only about six months before Western Electric produced a prototype of the No. 302.

PART 6

POSTWAR IN THE BELL SYSTEM

W orld War II had a much different impact on the Bell System than WWI (Brooks 1975, 208). During the first war, the Bell System was nationalized, although its operation was left with the existing management, and it supplied a limited amount of equipment to a small geographical area in eastern France. The second war was geographically dispersed, and sophisticated electrical apparatus was in more demand. The impact of WWII on the Bell System was substantial.

When WWII erupted, authorities recognized that the Bell System had a finely tuned organization that could serve the war effort just as it was. It was not nationalized. Bell Labs scientists were exempted from the draft. In secrecy, their main activity was the further development of British-originated radar equipment—not telephone design. Such equipment was subsequently manufactured by Western Electric, and by the end of the war the Bell System had provided about half of all radar units used.

The Bell System remained healthy, but its domestic customers suffered. In 1942 the War Production Board prohibited the manufacture of new telephones for nonmilitary use. Demand had sharply increased since the beginning of the war, so service declined. By the autumn of 1945, unfulfilled domestic requests for new telephones reached more than 2 million.

The war had hastened many technical advances, so there were plenty of improvements that could be made in telephones. There were then about 30 million Bell System telephones in service, and with a big backlog the Hawthorne Works was not able to handle the load (AT&T Archives 1992, 63). A new facility was in order, and it was built just inside today's beltway on the east side of Indianapolis. The Indianapolis Works opened in 1950 and the new telephone was soon made there (Taggart and Van Allen 2021). To help satisfy demands for a new telephone, serviceable No. 302 telephones were refurbished with a new plastic housing that made the old phones look like the new ones (Bell Labs 1956a, 313).

19

BELL LABS VERSUS AT&T HEADQUARTERS

Looking back at the design of the handset desk stand, tensions within the Bell System were apparent. Corporate headquarters heard complaints about appearance and enlisted outside artists with fancy ideas. They commissioned a handful of such artists, and even let their own employees enter the contest. The result was a number of elaborate Art Deco designs that could not be manufactured. We believe this contest was a half-hearted effort by Bell Labs, and this belief is based on two observations. First, approval for the oval D-type design was given at lightning speed such that serious consideration of an artist's model could not have been made prior to that decision. Second, in a 1930 letter, R. L. Jones says: "When we started with outside artists on the handset problem last spring, we did not consider that this was the way to do it" (Jones 1930).

Henry Dreyfuss probably entered corporate's contest, but then withdrew and sided with the laboratories. Dreyfuss's 1930 contract with AT&T headquarters was not renewed, and he subsequently received all his contracts from the labs. Bell Labs, which had more of an engineering aesthetic, won the argument and kept control of design of the No. 302 pre–World War II telephone.

By the end of the war, however, AT&T knew that they had an internationally famous designer on their hands with Dreyfuss. This was serendipitous. Nevertheless, shortly after work on their postwar telephone commenced, AT&T's directors snatched design control away from Bell Labs and gave it to Dreyfuss,

FIGURE 19.1 Sketch by Robert Hose of a proposed new telephone, 1946. (Illustration courtesy of Henry Dreyfuss Associates.)

ostensibly to avoid duplication between him and their in-house designer (Wallance 1956, 36). But Bell Labs had established their own world-class position by that time and were not about to let headquarters get away with this. As part of the deal, Bell Labs moved their principal designer into the Dreyfuss firm—along with sketches and ideas (e.g., fig. 19.1)—and ensured that he had hands-on responsibility for the new design.

Robert Haven Hose (1915–1977) was the designer who was moved from Bell Labs to the Dreyfuss firm in 1946. He was born in Sleepy Eye, Minnesota, graduated from the University of Minnesota in 1937, and earned a master's degree in architecture at MIT in 1940 (Flinchum and Meyer 2017, 180). At MIT, Hose met another architecture student, William Purcell, and the two men married sisters. Shortly before Hose was moved out of Bell Labs, Purcell was hired by Dreyfuss.

At Bell Labs, Hose and Dreyfuss would have known each other well since Dreyfuss had been consulting with Bell Labs since 1930. Hose was the head of the Bell Labs industrial design group that was disbanded on the condition that Hose be transferred to the Dreyfuss firm (Purcell 1982, 3). We believe there was some competition between Hose and Dreyfuss because we know that Dreyfuss was reluctant to hire Hose. Purcell (by then already in Dreyfuss's firm) told author Flinchum quite emphatically, imitating Dreyfuss: "Get him!" Thus Purcell (Hose's brother-in-law), rather than Dreyfuss himself, brought the new employee into the firm in 1946. Dreyfuss, however, ended up with full responsibility for the new telephone project and subsequent Bell telephone designs (quid pro quo?).

Hose was, in fact, a very talented designer. He became the second president of the Industrial Designers Society of America (IDSA) from 1967 to 1968, succeeding its first president, Dreyfuss. In addition to his work on Bell telephones, Hose became a design consultant to the Hoover Company and took that client with him when he left the Dreyfuss firm around 1961.

20

FINALLY, A
COMFORTABLE HANDSET

For insights into designing the new telephone, we turn to a summary compiled by Alvin Tilley, the principal design engineer at the Dreyfuss firm (Tilley 1953, 2): "The theme for the composition of receiver on stand in unison became a convex line lying transversely on top of a longer convex line. . . ."

This theme was eerily similar to the Kellogg No. 1000 telephone (see chapter 17). It is tempting to think that the No. 500 was a design that Robert Hose brought into the Dreyfuss firm when he joined it and therefore conclude that Hose should get sole credit for the design. That is not necessarily the case, however. By 1946, Henry Dreyfuss had been for 16 years consulting with the very group that Hose worked in at Bell Labs. Undoubtedly, this sketch includes ideas from both Hose and Dreyfuss. Patent assignments are not always reliable, but in this case they are probably revealing. The patent for the desk set was awarded to Dreyfuss and Hose jointly, whereas the patent for the handset was awarded solely to Dreyfuss (Dreyfuss 1948; Dreyfuss and Hose 1949).

Technical data from Bell Labs for the handset preceded information for the desk set to such an extent that they were almost designed separately, with only the theme for tying the forms together. The handset was designed first, and it is fair to say that Dreyfuss would have provided ideas for the handset to his staff in New York and asked them to prepare models.

Tilley describes designing the handle: "The convex curve in the handle of the handset was thoroughly justified. At one time, we tried a straight line, 'the shortest distance between two points being a straight line,' but we soon found the ends became very large and wasteful of material. This fact had been recognized in the existing and older designs." He goes on to say: "Assembly drawings were made to study the required volume, rough sketches were made for various designs, model drawings were prepared, and we immediately went into wood mock-ups of about eight designs" (Tilley 1953, 2). Those eight mock-ups are shown in figure 20.1.

Tilley further reports that the models were sitting on a desk one day when Dreyfuss walked in from out of town. He picked up one handset, held it to his ear, said it gave him "griptaphobia," then walked out. The staff immediately went to work and leveled out that design, making the rectangular cross-sectional area less fat and rounding the lower corners for greater comfort and increased finger

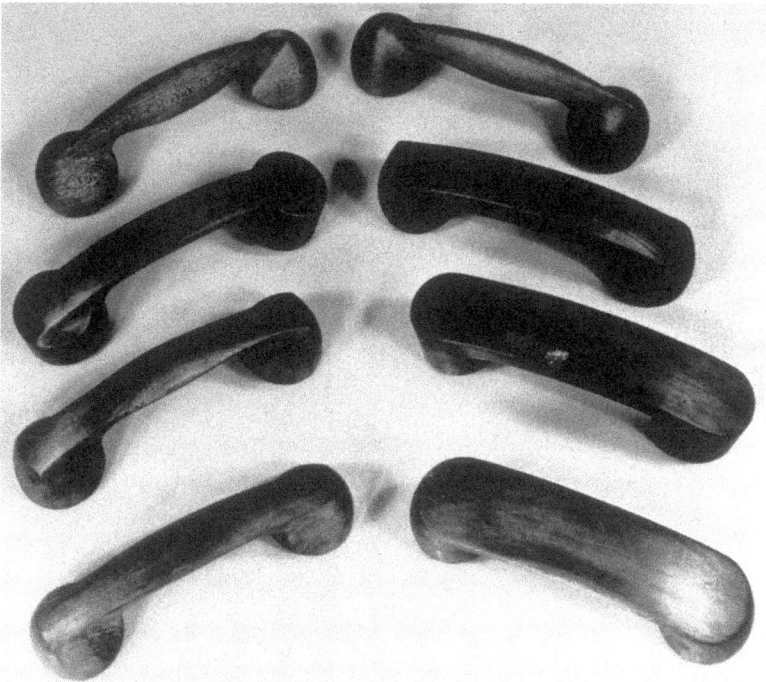

FIGURE 20.1 Dreyfuss firm handset mock-ups in wood, ca. 1946.
(Photo courtesy of Henry Dreyfuss Associates.)

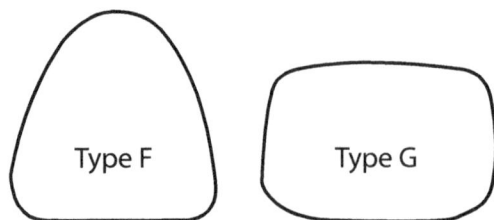

FIGURE 20.2 Mid-length cross sections of Western Electric's old Type F and new Type G handsets.

space, while maintaining the basic rectangular plan of the design. Thus, a less pure form was chosen that retained the original concept of a large, flat surface that the handset presented when placed in its cradle. The shape of the handle was important. The old triangular cross sections were awkward to hold and hard to service because they would not lie flat on a work surface. The flat-back shape of the new handset avoided those problems and has persisted (approximately) in all subsequent telephone handsets; this handle shape (fig. 20.2) is truly an undervalued feature.

From Tilley's notes we know that comfort was an intentional part of the design and that the rectangular cross section was less prone to turning in the hand. However, we can find nothing in the record that indicates the flat back would enable shouldering (fig. 20.3), which is a distinct advantage of the rectangular cross section. By the way, the handset in figure 20.3 is, in fact, from a later Automatic Electric Type 80 telephone (see chapter 23), but the handset dimensions and shapes are very similar to those of the Bell System Type G handset.

If we look at critical dimensions for the new handset, we can draw a significant conclusion. Figure 20.4 shows our measurements for this handset. The modal angle of 12½ degrees is about the same as in the Type E and F handsets (see table 11.1), but the modal distance is almost exactly ½ inch (13 mm) shorter than the previous handsets because this nominal reduction was requested by Bell Labs (Cobb 1952, 317). By the time of the new design—and probably long before—it was realized that it was not necessary to fit 99 percent of the measured heads; a slight tilt out of the plane of the ear would accommodate the longer heads. Further, sound pressure waves from a point source (roughly speaking, the mouth) are spherical and would be detected to the side of the mouth almost as well as the front. Dreyfuss was also free to reduce the tilt of the transmitter from 40 degrees

FIGURE 20.3 Young woman shouldering a modern handset. (Photo by Craig Robinson / Stock Unlimited.)

to 32½ degrees (7½ degrees is just a nominal three-quarters of 10 degrees) for more face clearance with no significant loss of performance because of the increased sensitivity of the new transmitter unit.

Figure 20.5 shows a human engineering drawing that helps visualize the head shape and clearances, and the dimensions that are shown are close to our measurements. Tilley created this drawing in 1965 for publicity purposes, long after the design was completed.

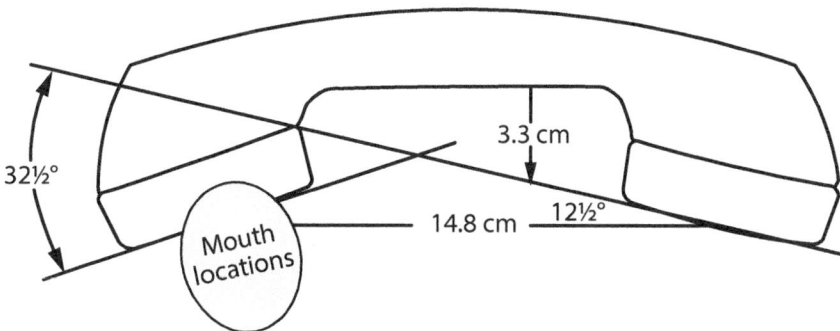

FIGURE 20.4 Outline and modal dimensions of Western Electric Type G handset.

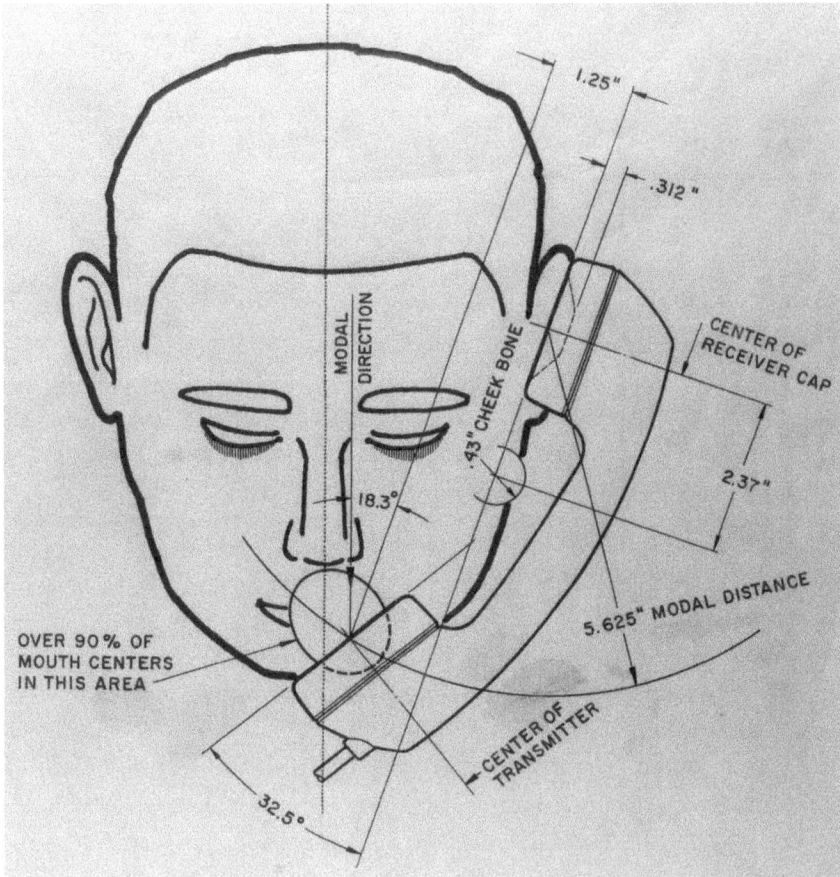

FIGURE 20.5 Alvin Tilley's drawing of the Western Electric Type G handset with modal dimensions. (Illustration courtesy of Henry Dreyfuss Associates.)

Figure 20.6 shows a similar drawing by Tilley with a data representation rather than dimensions (Bello 1954, 137). If you shade data cells in the old 1920s data presentation (see fig. 11.3) in groups of 20 percent of the total number, starting with the most frequent, you will reproduce exactly the patterns shown in figure 20.6. We can thus conclude that no new anthropometric measurements were made, notwithstanding the implications of these drawings. Except for the length, the dimensions of the new Type G handset were copied from the earlier handsets, and those dimensions were derived from the Bell System measurements of the

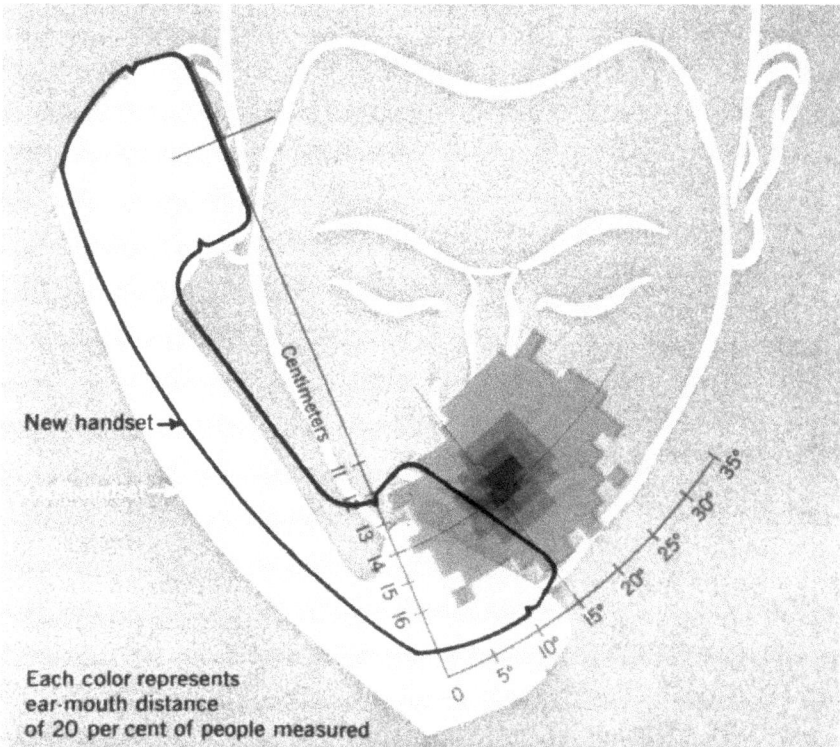

FIGURE 20.6 Alvin Tilley's drawing comparing the Western Electric Type G handset with the Type F handset and data. (Illustration courtesy of Henry Dreyfuss Associates.)

early 1920s. But the shape of the handle cross section was new, and this shape was significant beyond appearances.

The Type G handset was initially made of Bakelite, as were its predecessors, but that was soon changed to thermoplastic. The thermoplastic shell was hollow, so mild concern was again raised about howling. To counter this threat, a cotton ball was stuffed inside the end of the handle near the receiver element.

21

THE POSTWAR BELL SYSTEM STANDARD TELEPHONE

n the New York office, Robert Hose continued work on the base of the new No. 500 telephone. Alvin Tilley's colorful words describe the design concept: "All external surfaces were convex with optical cambers to reduce its apparent size. The predecessor was essentially an upper section of four concave walls superimposed over a rectangular base. Hence the original conception was the logical outcome of three of our axioms: 1) simplify form, 2) if it is circular make it square, and if it is square make it round, and 3) a design that is neither concave nor convex and without sex is the darndest thing!" (Tilley 1953, 1).

The new design provided an opportunity to utilize recent technical developments, but there were additional overt motives. One was to make maintenance simpler by mounting all components on a metal base and designing a housing as a mere shell that could be completely removed; previous sets had some components mounted awkwardly inside the housing. Another was to make the ringer loudness adjustable by the user to reduce the number of service calls for adjustment (Bredehoft 1951, 471). And the new technology made it possible to reduce complaints by adding an electronic equalizer that could compensate for the large variation in distance between customers and the central office—and hence variations in the resistance of the wires between the batteries and the telephone—thus decreasing performance variations (Meyer 2018, 180).

FIGURE 21.1 Western Electric No. 500 telephone introduced in 1949. (Photo by Matthew Gay, courtesy of the Gregg Museum of Art & Design, North Carolina State University.)

The new Western Electric No. 500 desk telephone was introduced in 1949 (fig. 21.1). Hose, who would work on the telephone desk set, moved only a short distance from Bell Labs to the Dreyfuss firm in New York City. Henry Dreyfuss, on the other hand, had moved his residence to South Pasadena, California, and opened another office there where he was close to some of his West Coast clients (e.g., Consolidated Vultee and later Lockheed Aircraft Company). Therefore, Hose and a local staff of about 30 employees were given the job of designing the new telephone in the New York office, with oversight by Dreyfuss of course.

The telephone's shapes were described by its designers, with just a little deprecation, as a shoe form (fig. 21.2) with a lumpy rectangle handle. Nevertheless, in 1959 the Bell System's No. 500 telephone was recognized as one of the ten "best designed products of modern times" in a poll of leading industrial designers (Bell Labs 1959, 151).

Tilley goes on to say that the actual design of the desk set commenced when Bell Labs provided measurements of the dial mechanism, ringer, electrical network, sound-level equalizer, and switch spring assembly. According to Tilley: "These units were put together in such a way as to reduce the overall height of the existing phone and lowering the dial and flattening its angle to improve the

FIGURE 21.2 Shoe form of Western Electric No. 500 telephone.

visibility and manual comfort of the user. Sketches were made, followed by accurate layout drawings. Full size models were studied in clay (plasticine)" (as shown in fig. 21.3). "When several designs appeared rational, they were cast in plaster and sculptured. Some were painted and equipped with mock components, such as handset, dials, cords, etc. to simulate the real thing" (Tilley 1953, 3).

Changes came sporadically, Tilley notes: "For example, it was found one day that three cents could be saved by using a certain wire on the windings with the result that we could lower the housing ⅛ of an inch [3 mm]. But some other details sometimes raised the housing; for example, to improve the bell tone, the diameter of the bells was increased" (Tilley 1953, 3). The result was that the design literally had its ups and downs.

After establishing the overall shape, Tilley notes that models were made with pockets cut out on both sides of the housing to receive the ends of the handset. Two problems resulted: first, the compact shoe form was degraded, and second, the demands on the telephone user were too exacting since the handset had to be lowered into these wells almost vertically. "As a compromise, we added two hook prongs in the rear and opened the wells in the back" (Tilley 1953, 4).

Throughout many months of meetings and transportation of models between the Dreyfuss office and Bell Labs, most design staff refused to accept the four-prong switch hooks that the engineers preferred. Tilley reports: "During one of the lull periods, a few of the staff casually tried modeling a four prong job and we

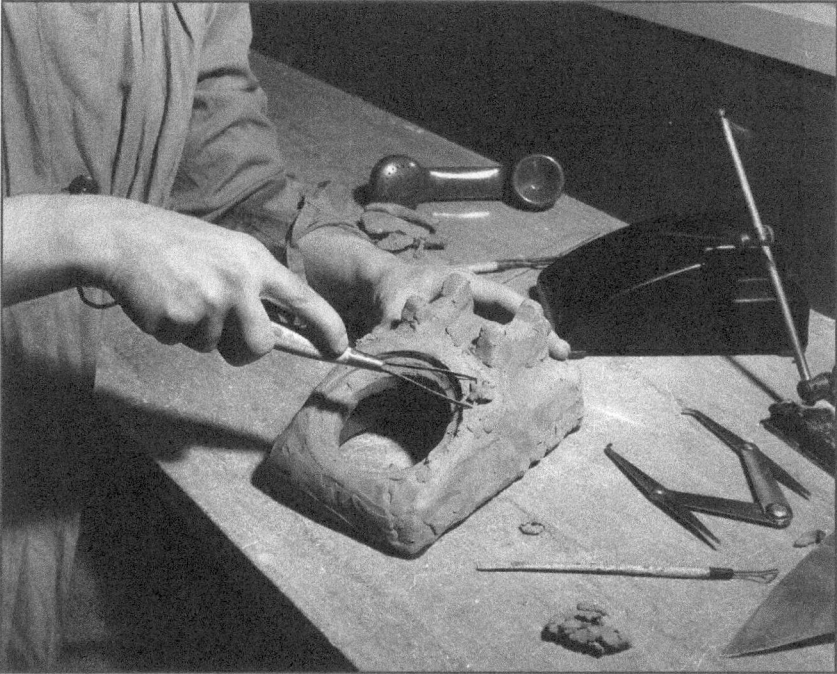

FIGURE 21.3 Sculpting model of Western Electric No. 500 telephone housing. (Photo courtesy of Henry Dreyfuss Associates.)

hit upon a happy coincidence of harmony between the front line of the prongs and the lines of the body cut-out. Heretofore, the front prongs always appeared detached. We had just finished smoothing the clay when Dreyfuss walked in on one of his quick inspections after a siege of out-of-town travels. He looked at the model, was interested, and asked his trusted model maker what he thought of the form. An affirmative answer was received, and, with a nod from HD, the battle of the prongs was over" (Tilley 1953, 4).

In addition to appearance, the desk set also included convenience features that contributed to its value as a good industrial design. Of course the loudness control on the ringer and the electronic equalizer—both mentioned above—were user conveniences as well as being features that reduced maintenance costs for the telephone company. But there were others.

A lifting feature, which would become a hallmark of desk telephones designed by the Dreyfuss firm, was incorporated as it had been for the No. 302 telephone. Like that earlier phone, this feature consisted of a cavity beneath the cradle (fig.

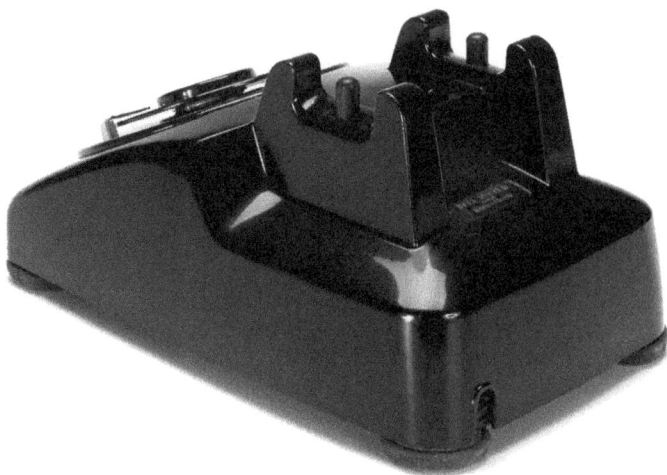

FIGURE 21.4 Rear view of the Western Electric No. 500 telephone showing the handhold cavity beneath the cradle. (Photo by Matthew Gay, courtesy of the Gregg Museum of Art & Design, North Carolina State University.)

21.4). One could grasp the phone, with or without the handset in place, and lift it easily with one hand. This was still an important feature because the post-war telephones were almost as heavy as their pre-war counterparts.

Renewed attention was given to the sound of the gongs, using the musical interval of a major third to produce a pleasing sound (Jenkins 1957, 350). The standard frequencies for the No. 500 set were fixed at 1,610 and 1,280 cycles per second and were produced by brass gongs (other materials were used to produce different tones for multi-phone office settings). These frequencies are lower than those in the No. 302 telephone, making the new ringer more audible, especially for older people with high-frequency hearing loss. A resonator shell (a Helmholtz resonator) was used to enhance the loudness of the fundamental frequencies relative to their overtones.

Another feature was the relocation of the dial numbers and letters from beneath the finger holes to outside the finger wheel (see fig. 21.1); Dreyfuss's favorite Futura typeface was chosen for the characters. The shift in position made these legible numbers visible from a wide angle and less subject to damage from dialing. The relocation of dial numbers was not new with the No. 500 design. This practice dated from the late 1920s when Gray, a pay phone manufacturer for the Bell System, used this feature to make numbers more visible in poorly lit pay

FIGURE 21.5 White aiming dots beneath the holes in a finger wheel.

phone booths. In those early pay phones, the background was white beneath a black finger wheel.

Testing of the new design prior to field trials showed that dialing time for the No. 500 was unfortunately slower than for the No. 302 (Black and Cunningham 1954, 26). Slower dialing was undesirable because it increased circuit-holding time, which tied up trunk lines. John Karlin, an experimental psychologist hired by Bell Labs, determined that people had difficulty telling when the finger wheel had stopped moving because it was black against a black background (Hanson 1983, 1576). He suggested placing white dots beneath the holes in the finger wheel, and these aiming dots solved the problem (fig. 21.5). Available reports do not say whether dialing with dots was actually faster than with the old dials, but the outboard numbers were not used on any future designs.

One appearance blemish of previous dials had been the tendency for their numbers and letters to wear off. To avoid this with the softer plastic number ring, a double injection molding process was used. The plastic ring was first molded with passageways for the white characters to be injected later. Thus the characters go all the way through the plastic ring and are not obscured by surface wear.

22

TRANSISTORS AND
PUSH-BUTTON DIALING

D uring the lifetime of the Bell System, five Nobel Prizes were won by Bell Labs staff and consultants (AT&T Archives 1992, 208). In 1956, three of the Bell Labs scientists received the Nobel Prize in physics for "investigations on semi-conductors and discovery of the transistor effect" (Bell Labs 1956b, 401). Following World War II, these investigations led to the invention of the transistor at Bell Labs. The impact of the transistor on all electronics was of course monumental. One of the early uses of the transistor was made at Bell Labs in a totally new push-button (Touch-Tone) method of dialing that incorporated electronic tone generators in the telephone sets, rather than impulse switches. The tone generators got their power directly from the telephone line.

Clear advantages were offered by this method of dialing. First, Touch-Tone dialing is much faster than rotary dialing. Second, so-called end-to-end signaling is possible; that is, the tone frequencies can be used after a connection has been made to transmit information (e.g., selecting from a menu). Further, it is easier electronically to handle alternating current signals within the voice frequency range than to process direct current pulses, which produce radio-frequency interference during the voltage ramps up and down, as with the rotary dial.

Because it was desired to use frequencies within the voice frequency range, a scheme had to be developed that would avoid false signaling (talk-off) due to music, speech, or background noise in the transmitter (Schenker 1960, 236). In

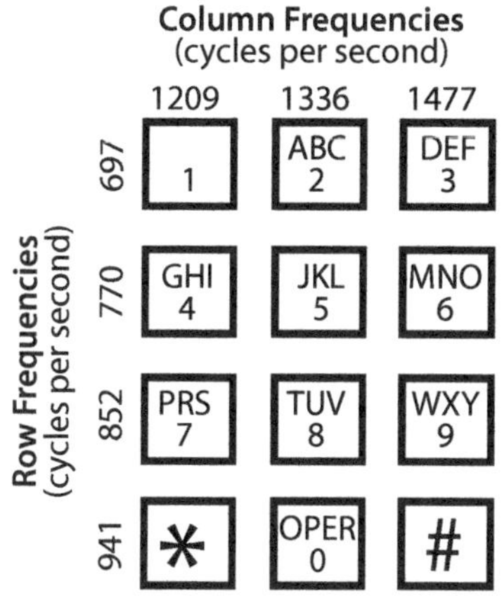

Column Frequencies
(cycles per second)

FIGURE 22.1 Tone frequencies for dual-tone multi-frequency (DTMF) dialing. (Illustration by John Voglewede.)

this scheme, eight tone frequencies were used, two at a time, in various combinations; this method was known as dual-tone multi-frequency (DTMF) dialing. When arranged in a 4-by-4 matrix (4 columns and 4 rows), these 8 frequencies would produce 16 unique frequency pairs. Generally, only 12 of these frequency pairs were used in a standard Touch-Tone dial (fig. 22.1), but all 16 of the frequency pairs were used by the US military in an automatic voice network called AUTOVON.

Having transistors and the engineering know-how to accomplish this was one thing; knowing the best way to lay out the buttons for the user was another. Should the buttons be round or square? Should they be arranged in circles, rows, or arrays? How about size and spacing? All of these parameters and more were investigated at Bell Labs.

John Karlin was head of the User Preference Research Department, and under his direction, they developed a testing system called Sibyl that could record dialing speeds, errors, preferences, and other parameters for a large number of test subjects (Hanson 1983, 1573). Bell Labs engineer Richard Deininger was given the job of determining the optimal parameters of the push-button dial, particularly the arrangement of the 10 buttons. Deininger used Sibyl with many volunteer test subjects to study 15 different button arrangements (e.g., fig. 22.2) along with button size, spacing, feel while depressing, and location of lettering (Deininger 1960, 999).

The net result was the now-familiar arrangement shown on the left in figure 22.2—a pattern that has become the international standard on objects as diverse as automatic teller machines (ATMs), vending machines, and medical equipment.

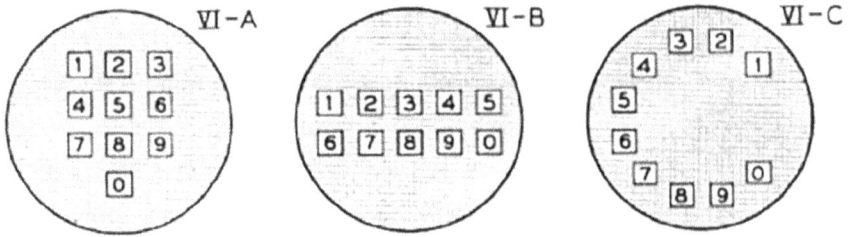

FIGURE 22.2 Three of 15 button arrangements tested for efficiency and comfort. (Illustration courtesy of the AT&T Archives and History Center.)

Notably though, calculator keypads continue the tradition of old mechanical cash registers that read from the bottom up rather than the top down.

A Touch-Tone version of the No. 500 desk phone with a keypad dial was introduced in 1963, and Donald Genaro was given design responsibility for integrating the keypad into the No. 500 desk stand. Donald Michael Genaro, a surviving associate of Henry Dreyfuss, was born in 1932 in Hoboken, New Jersey, and graduated from the Pratt Institute's world-renowned industrial design program (Flinchum and Meyer 2017, 186). Genaro's design modification was patented (fig. 22.3), although Dreyfuss put his own name on the 1962 patent application (Dreyfuss 1963).

Except for the Touch-Tone unit, which contained its transistorized tone-generating circuitry, the circuit and components mounted on the base remained almost unchanged. Therefore, Genaro left the back end of the 500 style alone and brought the front surface up to a flat 4-inch (10.2 cm)-square surface with ever-so-slightly bowed out edges. (Remember Tilley's words: "A design that is neither concave nor convex and without sex is the darndest thing!")

Like the rotary dial, the Touch-Tone keypad was mounted on a bracket on the base, yet the entire flat surface was also removable. The new Touch-Tone phone was introduced in 1963 with 10 buttons, but the star and pound key were added later. The large flat surface around the keypad made it easy to position the number card, to add a switch for two-line phones, and to add an indicator light for applications in private exchanges. With some stretching of this flat surface, a row of keys (switches) was also placed along the bottom to make multiline business phones.

By the way, the transition from pulse dialing to Touch-Tone dialing required large changes in infrastructure, and these changes did not take place readily in war-torn Europe. Some push-button dialing was introduced in European

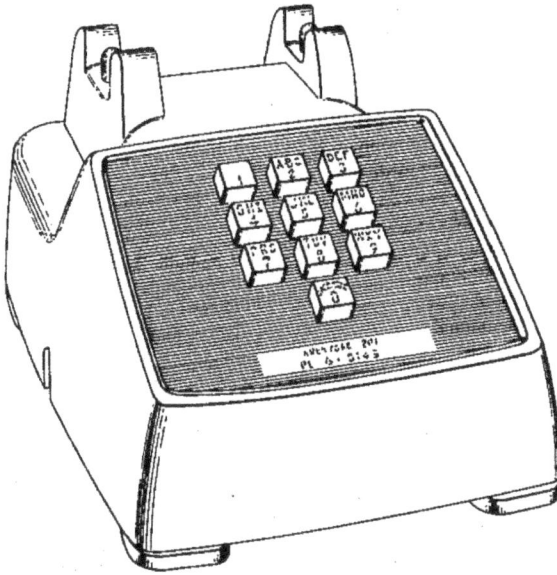

FIGURE 22.3 Drawing from the design patent for Genaro's Touch-Tone modification for a desk set. (Reprinted from Dreyfuss 1963.)

phones, but the early versions had pulse generators rather than dual-tone generators. Cell phones, which are rapidly displacing wired desk phones, continue to use this dialing arrangement, and the Bell System's dual-tone, multifrequency parameters have become an international standard (International Telecommunications Union 1988).

PART 7

OTHER VARIANTS AND DESIGNS

---

merican research and manufacturing facilities were intact after World War II, but similar facilities in Europe had been destroyed—and it took time to recover. Telephone service in Europe was slowly restored using prewar designs, and it was at least a decade before new designs emerged. Anyone who had read a magazine or watched a television program during that decade would have known about the American phone, so it is not surprising that the new European phones might be similar to the Bell System's No. 500 telephone. In America, competitors' phones were even more similar.

23

COPIES OF THE BELL SYSTEM STANDARD TELEPHONE

A fter three-quarters of a century of fierce competition between the Bell System and the independent companies—and, not incidentally, after two world wars and a raging Korean conflict—an event took place that led to much standardization in the American telephone industry. Heralded only by a two-paragraph note in *Telephony* magazine, a cross-licensing patent agreement was concluded on June 7, 1951, between Western Electric (for itself and AT&T) and International Telephone & Telegraph Company (ITT) and its subsidiaries (Telephony 1951a, 42). The agreement was intended "to strengthen the technical leadership of the free world" and it permitted each company to use the other's patents. A pending antitrust suit against Western Electric and AT&T filed by the US Department of Justice in 1949 might also have played a role in this matter.

KELLOGG AND STROMBERG-CARLSON

Two months later on August 9, 1951, ITT made a large purchase of Kellogg stock, eventually acquiring control of that company (Telephony 1951b, 42). Kellogg briefly manufactured a K-500 look-alike set, but soon Kellogg produced a completely standardized 500-type telephone like the Western Electric phone. The Kellogg telephone had a mixture of ITT and Kellogg markings on the housing, the handset, and the dial, but its parts were identical in appearance and interchangeable with those in the Western Electric 500-type telephone. Even the cotton ball that was used to prevent acoustic feedback was present in the Kellogg handset. Stromberg-Carlson also produced a No. 1543 look-alike phone, but likewise switched over in the 1960s to a standard 500-type telephone with interchangeable parts.

AUTOMATIC ELECTRIC

Automatic Electric similarly made a look-alike telephone and it was widely used, but they never switched to making a clone of the Western Electric phone. Automatic Electric's Type 80 telephone of 1954 (fig. 23.1) was very similar in appearance to the Western Electric No. 500 phone and contained a mixture of Automatic Electric and Western Electric technologies. The Type 80 had a network circuit in a package that looked quite different from the package for Western Electric's network, but the circuit inside was basically the same with only minor modifications. Automatic Electric retained its transmitter and receiver designs but housed them in a handset that was very similar to Western Electric's Type G handset. Automatic Electric later produced cheaper versions of the Type 80 telephone but never switched to the standard 500-type design.

FIGURE 23.1 Automatic Electric's Type 80 telephone, 1954. (Photo by Marketa Ebert.)

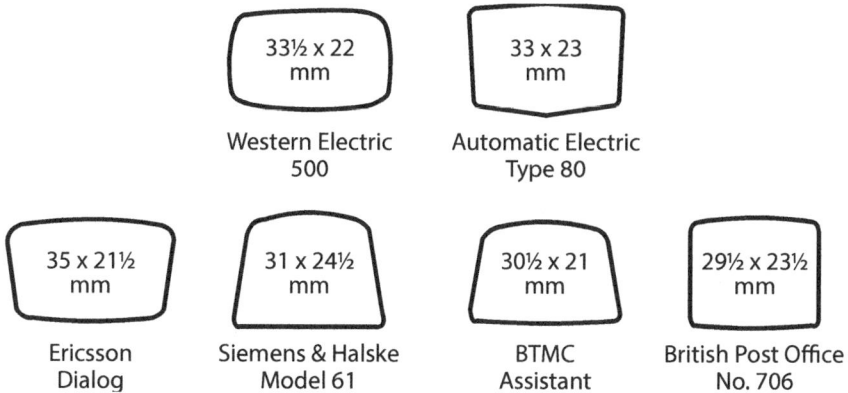

FIGURE 23.2 Approximate central cross sections of handles in modern handset telephones, with maximum width-times-thickness shown inside each outline.

The Automatic Electric Type 80 handset had almost identical modal dimensions as the Western Electric Type G. Table 23.1 presents a summary of post–World War II handset dimensions, and those of the Western Electric Type G handset are shown in boldface for ease of comparison with the dimensions of other handsets described below.

Figure 23.2 shows the approximate handset cross sections for the Automatic Electric Type 80 along with those for other similar postwar telephones. Like the Western Electric handset, the Type 80 handset shoulders well (see fig. 20.4). Although the overall dimensions of the Type 80's cross section are very close to those of the Western Electric No. 500, the shape is a bit different; the Automatic Electric handset and the European handsets were all cast in side-by-side molds rather than top-to-bottom molds like the Western Electric handset. Thus all of those handsets have a slight ridge down the center of their back surface.

BTMC ASSISTANT

The first of the new European phones was designed jointly by Bell Telephone Manufacturing Company (BTMC) and Standard Elektrik Lorenz (SEL), a German manufacturer of electronic devices that was also owned by ITT. Wolfgang Grüger and Henri Van Holst describe this design, the Assistant, in an article in a quarterly technical journal published by ITT (Grüger and Van Holst 1963).

Table 23.1 Summary of post-WWII handset measurements made by the authors*

HANDSET IDENTIFICATION	INTRODUCTION YEAR	LENGTH, δ (CM)	ANGLE, α (DEGREES)	TILT (DEGREES)	CLEARANCE (CM)
Western Electric Type G	**1949**	**14.8**	**12½**	**32½**	**3.3**
Western Electric Trimline	1965	14.3	13½	43	2.5
Western Electric Type K	1973	14.5	11½	30½	2.6
Western Electric Type R	1984	14.8	12	28	2.0
Automatic Electric Type 80	1954	14.9	12	31	2.9
BTMC Assistant	1957	15.3	13½	32½	2.8
British GPO No. 706	1959	15.9	19	47½	2.5
Siemens & Halske Model 61	1963	16.5	14	32	2.8
L.M. Ericsson Dialog	1964	15.5	13	37	3.1

*Approximate accuracy +0.1 cm, −0.3 cm, ±1 degree.

FIGURE 23.3 BTMC Assistant telephone, 1957.
(Photo by Remco Enthoven.)

The BTMC Assistant (fig. 23.3) was field-tested in 1956–1957 and introduced shortly thereafter. It replaced the Antwerp phone (see fig. 17.7) and was
one of the first thermoplastic telephones in Europe. Gray was a popular color
for telephones of the time, and the one shown here is gray. All of the working
components were mounted on a metal base, and the plastic shell was removable for servicing. Several circuits were provided, depending on the phone's usage, and one of those was equivalent to the Western Electric No. 500 circuit
with automatic equalization. The transmitter and receiver capsules were also
like the Western Electric units. The Assistant did not, however, have a carrying handle, although one was added to later spin-off designs in Europe. Versions of the Assistant were also produced by SEL in Germany and by manufacturers all over the world.

Of special interest in all of these telephones is the handset design, and in 1932,
the Bell System had published head-measurement results that were made a decade earlier (see chapter 11). Two years after that report was issued, Hans-Joachim
Lurk at Siemens & Halske in Germany reported 5,000 head-measurement results that had been made with a different apparatus (Lurk 1934). More recently
Grüger and Van Holst refer to Lurk's paper as well as to the Bell System publication. For our purposes, Lurk's article is largely a red herring, but it provides

important clues to other head-measurement data that are relevant and not well reported.

In particular, Lurk's paper mentions an earlier Siemens & Halske set of 1,000 head measurements and points out that they were taken with the straightforward Bell System apparatus rather than Lurk's new complicated apparatus. Lurk's paper shows that the earlier Siemens & Halske dataset differs by about 5 degrees in α when compared with the Bell System data, and both of those datasets were measured with the same Bell System apparatus. The difference between these two datasets is probably significant and likely due to differences in the population of subjects measured in the two studies. The Bell System had selected subjects for their measurements that included both sexes and various races in about the proportion indicated by the 1920 US census. Siemens & Halske's subjects were employees and family members, and this population might have been different in Hitler's 1933 Germany.

In discussing the design of the BTMC Assistant, Grüger and Van Holst chose to use the earlier Siemens & Halske 1,000-measurement dataset, based on the Bell System apparatus, and plotted those results along with an outline of the Assistant's new handset (Grüger and Van Holst 1963, fig. 7). As before, the angle between the plane of the ear and the center of the mouth was α, and the modal distance between the center of the ear and the center of the mouth was δ. We have added two oval-shaped outlines to their figure in our figure 23.4, and these outlines enclose 90 percent of the earlier Bell System data and of the 1,000-measurement Siemens & Halske data, respectively. Notice that the BTMC Assistant handset was not a close fit to the Siemens & Halske data. The handset was a closer fit to the Bell System data, but it might have been an even better fit if its length had been reduced ½ inch (13 mm) as done on the postwar Western Electric Type G handset (see table 23.1).

BRITISH GENERAL POST OFFICE NO. 706

In Britain, telephones and service were provided by the General Post Office (GPO), and until the No. 706 (fig. 23.5) was introduced in 1959 the official telephone was a version of the Ericsson 1931 model (see fig. 17.5). The new telephone was designed jointly by Ericsson Telephones Ltd., the GPO Engineering Department, and the Council of Industrial Design, and the phone was made by more than a half dozen different manufacturers (Spencer and Wilson 1959, 12).

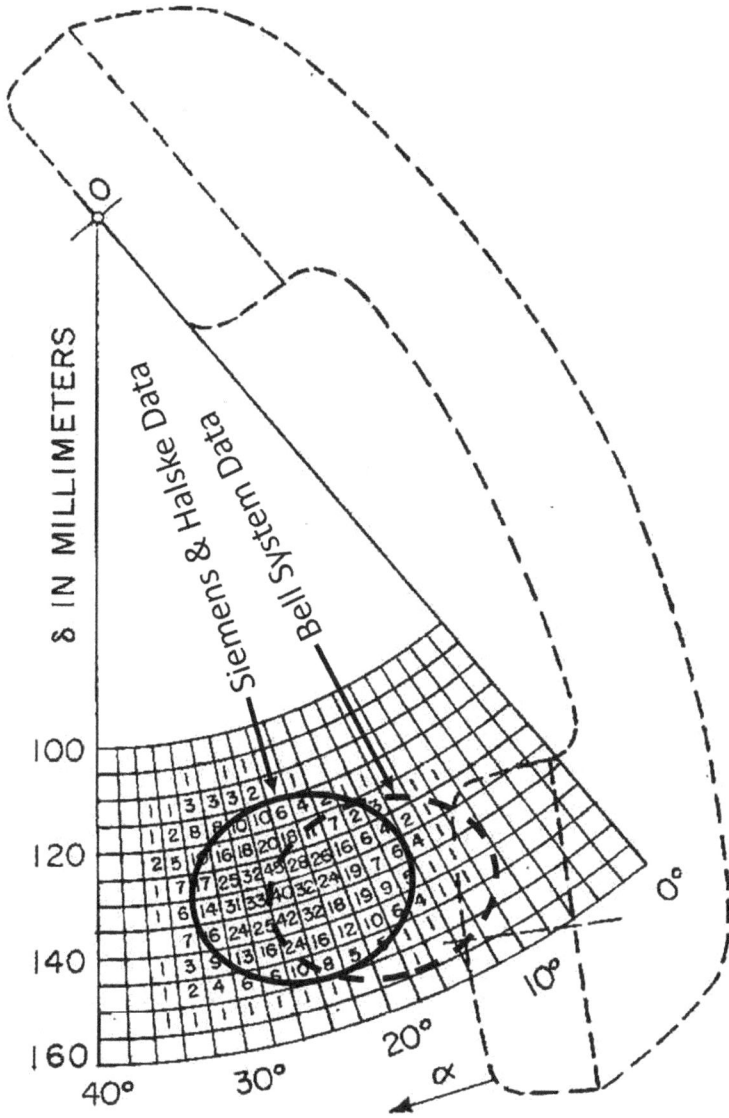

FIGURE 23.4 Drawing of handset for BTMC Assistant telephone with data outlines added. (Adapted from Grüger and Van Holst 1963, fig. 7.)

FIGURE 23.5 British General Post Office No. 706 telephone, 1958. (Photo courtesy of BT Group Archives.)

The thermoplastic housing was a removable shell, and all components were mounted on the base. An optional carrying handle could be mounted under the cradle. The rotary dial had numbers under the finger holes as well as outboard numbers on an external ring for ease of viewing. The electrical circuit incorporated a regulator that was sensitive to the line current.

As shown in table 23.1, the handset had a modal length of 15.9 cm. This was longer than the GPO version of the prewar Ericsson phone that used the Neophone handset (15.0 cm; see table 11.1). The cross section of the handset was rather narrow, however, such that the handset did not shoulder well.

SIEMENS & HALSKE MODEL 61

The Post Office in West Germany also provided telephones and service, and manufacturing resumed in 1948 on a very slightly modified prewar design (see fig. 17.1) (Arbenz 2009, 180, 197). That phone was replaced in 1963 by a new model (fig. 23.6) that was initially available only in gray, so the phone became known as the "Gray Mouse." The Siemens & Halske Model 61 (after 1966 just called Siemens Model 61) was a further development of the BTMC Assistant and had a

FIGURE 23.6 Siemens & Halske Model 61 telephone, 1963.
(Photo by Dietrich Arbenz.)

carrying grip, removable plastic housing, adjustable ringer loudness, and all components mounted on a plastic base.

The handset again had, more or less, the same dimensions as its predecessors (see tables 11.1 and 23.1), exhibiting a stubborn reluctance to a change in modal dimensions—about 2 cm longer than the Western Electric Type G. The handle's cross section was rather narrow, making the handset somewhat difficult to shoulder. The Model 61 was manufactured in Argentina and Ireland as well as in West Germany.

ERICSSON DIALOG

Sweden did not suffer the extensive destruction that occurred elsewhere in Europe during WWII. Thus, shortly after the war, Ericsson was able to invest heavily in a new design but chose to work on a radical design called Ericofon (see fig. 27.8). This effort delayed work on a new standard telephone. Thus the old Ericsson 1931 model (see fig. 17.5) remained the most common phone in Sweden until 1964.

A description of the new standard telephone, the Dialog (fig. 23.7), was given in the 1964 *Ericsson Review* (Kåell 1964). All of the components of the Dialog were mounted on a steel base, and the housing was a plastic shell that could be easily removed for servicing. A recess at the rear of the body provided a

FIGURE 23.7 Ericsson Dialog telephone, 1964. (Photo by Remco Enthoven.)

convenient grip for carrying the phone. Although Ericsson retained its traditional dial numbering configuration, a rim was added outside of the dial for extra outboard digits or letters should they be needed. Ringer loudness was adjustable by the user, and the electrical circuit automatically regulated the transmission level to compensate for different line lengths.

The handset was shorter than that of its predecessor, but it was still significantly longer than that of the Western Electric Type G. Nevertheless, the handle was quite wide such that the handset shouldered well. Ericsson had an extensive export business that included factories in Brazil, Australia, and Mexico as well as in the Nordic countries.

In summary, all of the telephones in this chapter had housings with a smooth shoe form that looked nothing like their square-shaped predecessors. And all of these post-WWII telephones had handset handles that were basically lumpy rectangles in cross section, in contrast to the triangular cross sections on the phones they replaced. Further, all had removable plastic shells with components mounted on a base plate, all had adjustable ringers, and almost all had carrying handles. These features had been notable developments in the Western Electric No. 500 telephone. Except for the two American manufacturers who cloned the No. 500, there is no question that the other producers did their own engineering and determination of dimensions and final shapes, as is apparent from records of that work. Nevertheless, from an industrial design point of view, the design concepts were apparently copied because their features are too similar to the No. 500 to be coincidental. It is thus fair to say that the design of the Western Electric No. 500 telephone was copied around the world.

24

THE BELL SYSTEM'S
BEDROOM TELEPHONE

By the mid-1950s, Bell Labs had a new marketing division and a recent corporate decision to pursue a full-scale, full-time program of selling its products and services—and devising new ones to sell (Brown 1966, 13). The first result from this program was an extension phone, designed primarily for use in a bedroom. All earlier Bell telephones had been designed to incorporate technical advances that improved performance or lowered costs, but no such features were incorporated in this telephone. The design was purely for marketing purposes. This would also be true of the later dial-in-handset telephone. There were additional designs that we would call whimsical or innovative, but not necessarily good industrial designs, and these (mentioned below) used existing technology as well.

The Bell System introduced its Princess extension phone (fig. 24.1) in five colors in 1959. With its decorator appearance and usual bedroom location, black was not its favored color, and a black Princess phone was not introduced until later. The pink Princess was popular and the Princess led the way with color, but the dominant color for all Bell System phones subsequently became almond (a light beige or dark ivory) rather than black.

Although the Princess phone was a radical departure from traditional telephone designs of purely functional instruments, its concept was not entirely new. The failed 1929 design competition produced one interesting design that was

promise a *Princess*

Let this lovely little Princess phone light up the season for someone you love. She'll welcome the convenience of an extension phone, and the added compliment of good taste.

FIGURE 24.1 Magazine advertisement for the Princess telephone introduced in 1959.

similar to the Princess. In that competition, Gustav Jensen, a well-known independent designer, had proposed an oblong Art Deco design with the handset straddling the dial (see fig. 14.2). Three later patents with this configuration were cited in the eventual patent for the Princess design, but none was as similar in appearance to the Princess as was Jensen's design.

The Princess telephone used the same Type G handset as the No. 500 phone, but the desk set had a distinctly different footprint and, therefore, required reconfiguring the interior components (notice that the dial no longer had outboard numbers). The component configuration was done by Bell Labs in New Jersey. Exterior design of the Princess desk set was thus a job for the Dreyfuss firm's nearby New York office.

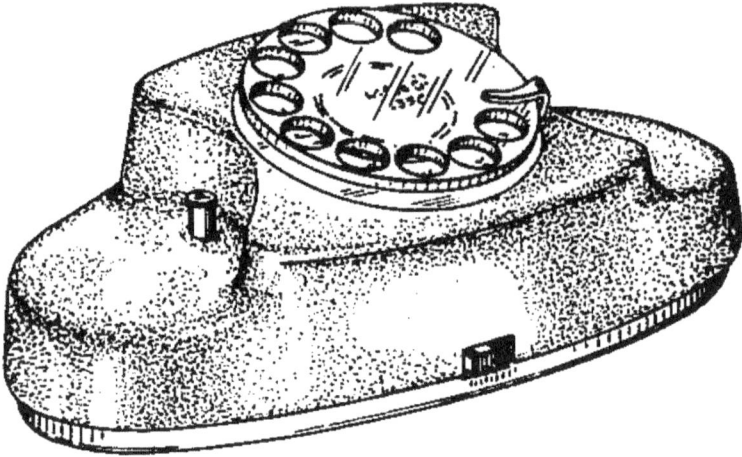

FIGURE 24.2 Drawing from the design patent for the Princess telephone. (Reprinted from Burlin and Hose 1958.)

By the mid-1950s, James Burlin was managing the day-to-day activities of the Bell Labs account and would have handled the many meetings and interactions with the laboratories. A patent for the ornamental design of this desk set was issued to Burlin and Robert Hose jointly, and one of the drawings from their patent is shown in figure 24.2 (Burlin and Hose 1958). We believe, however, that Hose was the principal designer of the Princess as Burlin was primarily a manager. Henry Dreyfuss's name is not on the 1956 design patent; this is unusual and perhaps significant. During the time period that design work on the Princess was being done, Dreyfuss had just built a house in California and opened a new office there; written his autobiography that was published in 1955; and designed a successful interior for Lockheed Aircraft's Super Constellation as well as the interior of Eisenhower's similar Air Force One (Flinchum and Meyer 2022, 53). It is thus likely that Dreyfuss took an uncharacteristic hands-off role in the Princess design—leaving it to Hose, the former Bell Labs designer.

By the autumn of 1957, market trials were underway in Illinois and Pennsylvania with prototypes that looked exactly like the patent drawing (AT&T 1957, 56). On these sets, as on the patent drawing, there was an overhanging lip just below the dial. This lip and a similar one in the rear were intentional features of the design to permit the desk set to be picked up with one hand straddling the handset—a signature feature of the Dreyfuss firm's desk telephones. The overall shape of the housing was pyramidal (i.e., bigger at the bottom) such that it could be molded in a single piece, but the overhanging lip prevented this. Consequently, the housings for field trial sets in 1957 were molded in two pieces that were glued together around this lip (McFadden 1994). This arrangement created an extra step in the manufacturing process.

Donald Genaro, who was often Dreyfuss's problem solver, was given the job of simplifying the molding process. Genaro recognized that a way to gain a lip in a single piece was to use a slide in the die to create a small area with a reversed draft and an undercut. The only minor appearance drawback to this approach was the witness lines that were created by the slide; these lines can be seen faintly in figure 24.1. To accomplish the die modification, Genaro worked with a small Bell Labs group that was located in the Western Electric factory complex in Indianapolis.

Another convenience feature of the Princess was an internal light that doubled as a night light and a dial light. When the handset was lifted, the soft night light would brighten to illuminate the dial for nighttime dialing. A sliding switch was provided so the night light could be turned off, but the dial would still light up when the handset was lifted. That switch is visible in the patent drawing (fig. 24.2) and the field-trial set, but for production the switch was moved to the rear edge of the metal base, thus further simplifying the housing. The internal light was incandescent and required power from a baseboard transformer via an extra pair of wires in the line cord.

For several years there was no ringer in the Princess desk set and a half-pound (0.2 kg) lead weight was added to keep the phone from moving while being dialed; this was similar to the situation with the 1920s desk stands. Eventually a small ringer was installed in the set. As with the No. 500 telephone, a Touch-Tone version of the Princess desk set was produced. Genaro again made the design modifications to accommodate the Touch-Tone keypad. By mid-1965, more than 4 million Princess telephones were in service, and the Princess (in pink, of course) is in the collection of the Smithsonian's National Museum of American History (Brown 1966, 14).

Since the industry had moved toward standardization following the introduction of the No. 500 telephone, the Princess phone was also copied by other manufacturers. But the name Princess was registered as a trademark by the Bell System, so other names were used. Kellogg's Cinderella phone had parts that were interchangeable with the Princess; Stromberg-Carlson's Petite phone looked just like the Princess; and Automatic Electric's Starlite phone was similar in appearance.

25

THE DIAL-IN-HANDSET TELEPHONE

T he familiar Trimline telephone of 1965 (fig. 25.1) had a long and tortuous period of development (Krumreich and Mosing 1966; Sentenne 1965). The concept at Bell Labs began with the lineman's test set of 1939, which had a rotary dial and all the telephone circuitry enclosed in a bulky handset. Start-ing in the early 1950s, five different dial-in-handset designs were developed sufficiently to be field-tested: Demitasse, Shmoo, Contour, Trimline I, and Trimline II. Two of those preliminary designs as well as the final production version of the dial-in-handset phone, simply called the Trimline, are described below.

FIGURE 25.1 Trimline telephone, introduced in 1965.

THE SHMOO

Dreyfuss's first dial-in-handset design (fig. 25.2) was called Shmoo because its shape resembled the bulbous cartoon character with that

FIGURE 25.2 Western Electric dial-in-handset Shmoo
telephone, 1959. (Photo by Paul Fassbender.)

name created in 1948 by Al Capp for the *Li'l Abner* comic strip. This telephone design was Henry Dreyfuss's personal creation, and it was patented in 1959 (Dreyfuss et al. 1959). The patent was shared with Robert Hose and James Burlin
presumably because they managed Dreyfuss's New York office and the many interactions with nearby Bell Labs that were needed to engineer the internal components for the phone. The Shmoo telephone used the same dial as the Princess.

Field trials with the Shmoo phone were first performed in New Jersey in 1959.
Although the dial-in-handset format was twice as popular as the No. 500 set,
many subjects complained that the handset was too bulky to hold comfortably
(Krumreich and Mosing 1966, 13).

TRIMLINE I

A small space-saver dial was engineered by Charles F. Mattke in 1959 at Bell
Labs (Hanson 1983, 1578). The smaller dial eliminated the gap that can be seen
between 0 and 1 on the Shmoo (fig. 25.2). It used a movable finger stop to facili-
tate removal of the gap. Although often unrecognized, the concept for this small

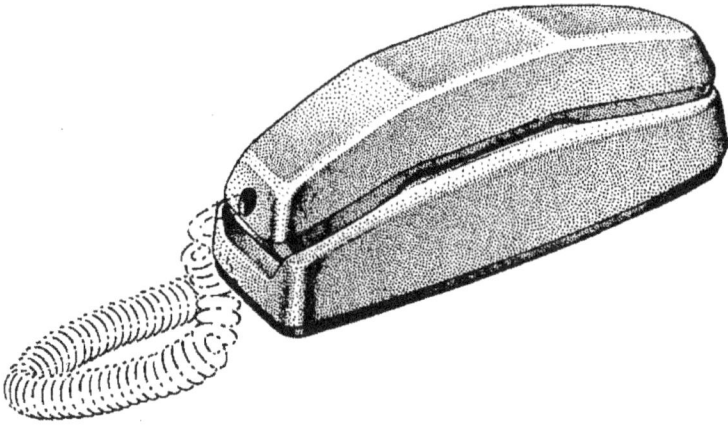

FIGURE 25.3 Drawing from the design patent for the Trimline I telephone. (Reprinted from Mosing 1961.)

dial was not new but had been developed in the late 1920s by Siemens & Halske (see chapter 27). The small dial was used by Lionel Mosing to fashion a slimmer dial-in-handset telephone of a more angular form that Bell Labs called the Trimline, later called Trimline I (fig. 25.3) to distinguish it from the final Trimline. Dreyfuss was unaware that a group within Bell Labs had created the Trimline I.

TRIMLINE

Donald Genaro says that Dreyfuss was deeply troubled by the extremely poor response to his Shmoo concept and Bell Lab's clandestine development of their own Trimline I dial-in-handset telephone. Eventually Bell Labs assured Dreyfuss that his firm would be responsible for the dial-in-handset phone. Hose, the former Bell Labs designer, had recently left the Dreyfuss firm, and Dreyfuss subsequently gave the final design work to Genaro in the New York office. Genaro thought that the Trimline I design was unrefined and concluded that nothing was salvageable from the previous dial-in-handset designs, so Genaro said they started with a clean slate. It must be said, however, that Genaro's clean slate did not start from scratch but rather with knowledge that the slim shape of Trimline I had been field-tested successfully and must be the configuration of Genaro's ultimate design.

Genaro made a preliminary design and then worked with the Bell Labs team at their nearby Holmdel, New Jersey, facility. He recalled the days in 1963 as being

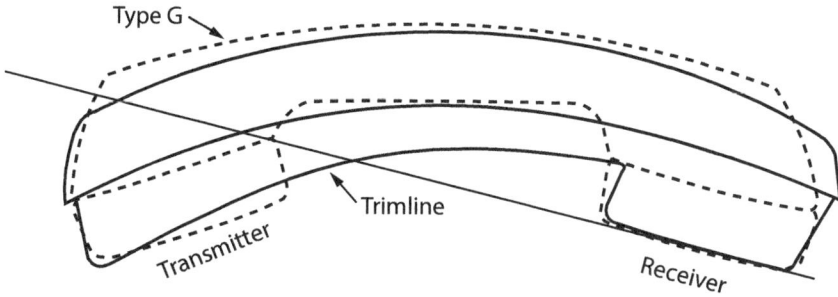

FIGURE 25.4 Comparison of the outlines of the Type G and Trimline handsets.

a series of running changes in components that had some effects here and there on the design. He continued making small adjustments, especially involving the details of the base's contours around the pockets that cradled the handset. He wanted the user to be able to replace the handset with ease and not have to carefully place it in its nest. Genaro recalled that the No. 500 set excelled in that respect: "You could almost throw the handset in the general area of its cradle and it would home in. Receiver off hook (ROH) was to be avoided; the set should appear hung up if it indeed was, and if it wasn't the set should look in disarray" (Donald Genaro, pers. comm. August 8, 2016).

In spite of all these changes, the modal dimensions of the Trimline are very similar to those of the standard Type G handset (fig. 25.4, and see table 23.1). Patents for the Trimline were issued to Dreyfuss and Bell Labs employees Mosing and Robert Prescott, but Genaro's name was not mentioned (Dreyfuss et al. 1965a, 1965b).

Whom should we credit for designing the Trimline? It's a fair question. Genaro points out that every line, contour, and profile down to the smallest appearance detail was his doing. Thus Genaro could be recognized as the creator of the only Dreyfuss-related design in the collection of the Museum of Modern Art, which does credit Genaro for this sculpture.

On the other hand, Genero was frequently the problem-solver, cleaning up various designs. He had indented panels in the Princess to provide a handhold feature, and he had reshaped the front of the No. 500 and the Princess to accommodate the Touch-Tone dial. From Dreyfuss's point of view, it might have been reasonable to see Genaro's work as merely modifying Dreyfuss's Shmoo or the Bell Labs Trimline I to create the production Trimline.

The changes Genaro made might have seemed minor, but from an artistic point of view the last 10 percent of a design is very important. The smooth lines

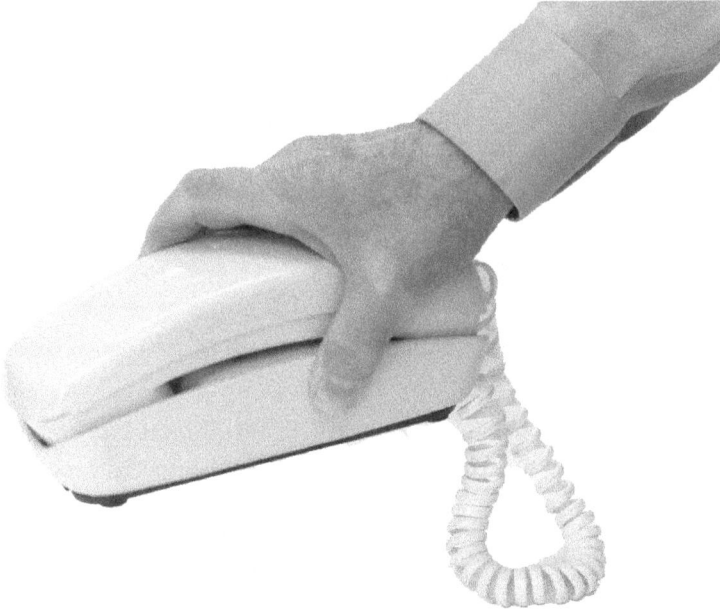

FIGURE 25.5 The reverse draft of the base makes grasping
and lifting possible. (Photo by Ted Mueller.)

of the handset without a defined transmitter, the reverse draft of the base with
its rolling under, and other subtle features gave the production Trimline the artistic qualities that made this telephone a favored museum piece. Because our
interest is in material design and because the patents in question are design patents, we believe that the design patents inappropriately omitted Genaro's name.

A characteristic of all Dreyfuss-related desk set telephone designs was a feature allowing the user to pick up the phone with one hand straddling the handset (fig. 25.5). Grasping of the Trimline desk set was made possible by the reverse
draft (inward-sloping sides) of the base, which also contributed to the perception
that the phone was visually lighter and had a high degree of finish and subtlety.

The Trimline also had a dial light but no night light. As in the Princess, the
light was originally an incandescent bulb that required power from a baseboard
transformer via an extra pair of wires in the line cord and in the handset cord.
The five wires in the handset cord made it uncomfortably stiffer than the usual
handset cord, and this also added cost. A second generation of production Trim-
lines had light-emitting diodes (LEDs) that operated off of telephone line power
rather than incandescent lights that needed external power; this eliminated the

need for a transformer and permitted use of the same lightweight handset cord as on other phones (Meyer 2018, 109, 195). This incorporation of LEDs required a change in the telephone circuit and thus could not have been done by a designer alone.

Like the Princess telephone, the Trimline was copied by other manufacturers using various trade names. Many parts were interchangeable between the Bell System's Trimline, Kellogg's Trendline, and Stromberg-Carlson's Slenderet— and the three phones were almost indistinguishable in appearance. Automatic Electric's Styleline was a little different but followed the same format. Several other dial-in-handset telephones were designed outside America (see figs. 27.2, 27.7, and 27.8), but none gained traction like the Trimline. And later developments of cordless and cell phones began their evolution in forms like the Trimline.

26

WHIMSICAL
BELL SYSTEM DESIGNS

I n 1968, a tri-company project team within the Bell System was formed con-
sisting of AT&T marketing and customer services specialists, Western Electric merchandising and manufacturing experts, and Bell Labs engineers (Hall
1975, 236). The team's purpose was to develop "antique/decorator" concepts into
new products and find ways to compress the timeline for products to reach the
market. The design requirements specified by the team were straightforward: the
telephones must be unusually attractive; use standard, available components; and
appeal to a broad range of customers interested in decorator-style telephones,
from the low-price mass market to the high-fashion, exclusive market.

Bell Labs engineers called on Dreyfuss's firm to develop a series of new-style
telephones in unique shapes, materials, and colors. Other designers were later
brought into the program, and the new products were all called Design Line telephones. For these phones, the customer would purchase a Bell System–approved
housing from a retail outlet, but the Bell System would retain ownership of the
components necessary to make it a working telephone set.

The Dreyfuss firm provided the first 12 designs of the Design Line series, and
they went into production beginning in 1975. The total number of eventual designs and variations in the series reached more than 50—eventually including
candlesticks, Disney characters, Snoopy, and Kermit the Frog. The Dreyfuss designers did not consider these phones to be a high-water mark of consumer taste

FIGURE 26.1 Western Electric Stowaway Design Line telephone, 1975.

as they critiqued the lineup from candlesticks to phones in the likeness of mice and dogs (Mickey and Snoopy). Faced with this whimsical product line, Donald Genaro described the firm's efforts as trying to make the designs submitted by their office as sensible as possible.

Genaro had oversight of all the Dreyfuss firm's Design Line telephones because Henry Dreyfuss had left the firm by this time. One of the first in the series was the Stowaway (fig. 26.1) that was in fact designed by Genaro and utilized a new Type K handset. Several cabinet styles were available, this one being in quarter-sawn oak.

The Type K handset was also designed by Genaro, with help from John McGarvey, and it is shown with the earlier Type G handset in figure 26.2 (Genaro and McGarvey 1974). Genaro reports that on a visit to Indianapolis while watching the molding and assembly of the Type G handset, he thought, "There's got to be a better, faster and less expensive way (and better looking). I might have been seen as attacking mom & apple pie, but with many of the sets I was beginning to work on (mostly business sets), the old dumbbell look of the G and round pockets troubled me" (Donald Genaro, pers. comm. July 31, 2016).

FIGURE 26.2 Comparison of the Type G (left) and Type K (right) handsets. (Photo by Ted Mueller.)

The Type G handset, with its rounded ends, blended well with the rotary dial No. 500 and the oval shape of the Princess-type telephones. But rounded shapes were dated. In fact, the No. 500 telephone did not have rounded pockets as did the Princess—those areas of the base had been scooped out in the No. 500 design—and the Type K handset looked great on the squared-off No. 500 Touch-Tone variation. But in 1973 when the Type K was developed, the No. 500–type telephones were in their late stages of production, and the Bell System did not backfit these earlier phones with the Type K handset. The Design Line phones gave Genaro an opportunity to use it.

Our measurements (fig. 26.3) show that the modal distance and angles on the Type K handset were almost identical to the Type G handset (see table 23.1), although the handle cross section was slightly different. The original method of fastening the top and bottom plastic shells together sometimes led to squeaking when users changed their grip, but this was later fixed by simply gluing the two halves together. The demarcation between the handset moldings was still proudly on display in the Type K handset, just as were the milled grooves in the handset of the old 1930 No. 202 telephone.

For the Design Line series, Genaro also designed the Sculptura phone (fig. 26.4) that is found in some museums. Genaro said that the Sculptura arose from his desire to have the handset become integral to the overall design rather than its typical placement as an add-on. This concept was somewhat like the Trimline that Genaro had shaped earlier, where the handset and base together comprised a sculpture.

Another notable member of the Design Line series was the Exeter (fig. 26.5) that was designed by John McGarvey and Gordon Sylvester. It was the most

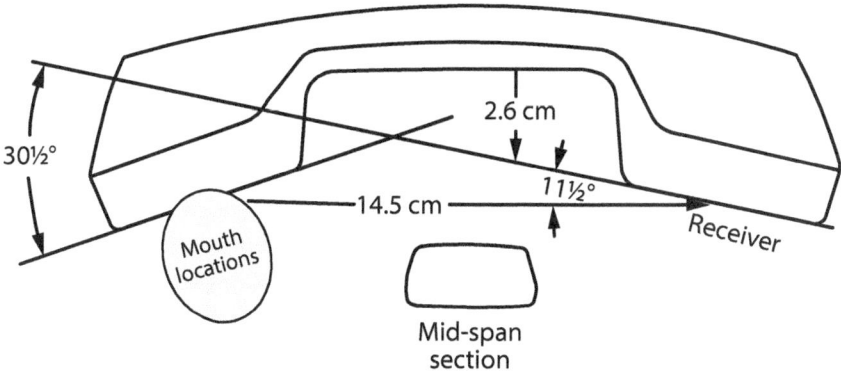

FIGURE 26.3 Outline and modal dimensions of the Type K handset.

practical of the offerings and exhibited a format like the business sets to come, for example the AT&T No. 700 (fig. 26.6). Many variations of this design were made by Genaro by enlarging and rearranging the flat panel.

FIGURE 26.4 Western Electric Sculptura Design Line telephone, 1975. (Photo by Matthew Gay, courtesy of the Gregg Museum of Art & Design, North Carolina State University.)

FIGURE 26.5 Western Electric Exeter Design Line telephone, 1975.

FIGURE 26.6 AT&T No. 700 business set, 1984. (Photo by Paul Fassbender.)

27

INNOVATIVE INTERNATIONAL DESIGNS

Over the years there have been some unusual designs that can't be ignored even though they might not be good industrial designs. Below are some of those. These innovative telephones contained some new features that later made their way into more conventional phones, so they are important as well as interesting.

FULD & COMPANY FRANKFURT TELEPHONE

At about the same time that Siemens & Halske was developing their new telephone in the mid-1920s, Fuld & Company also developed a new telephone (fig. 27.1, bottom image) (H. Fuld & Co. 1932, A27). The new phone, which was called the Frankfurt telephone, included a totally integrated handset cradle that is found in much later telephones. The German Post Office, however, did not select the Fuld & Company design, selecting instead the Siemens & Halske design (see fig. 17.1) as the standard telephone for Germany, so the Frankfurt phone did not have a large ready-made market. Fuld & Company nevertheless found an adequate market for the new telephone in their hometown of Frankfurt in a housing project that would have 12,000 apartments: the New Frankfurt project. It was initiated in 1925 amid political chaos in Germany and Ernst May, the city planner of Frankfurt, was named as the project's general manager.

FIGURE 27.1 Journal cover showing the 1929 Fuld & Company Frankfurt telephone (bottom image). (Reprinted from *Die Form, Zeitschrift für Gestaltende Arbeit*, April 15, 1930.)

Table 27.1 Prizes in Fuld & Company 1927 design competition

PRIZE	RECIPIENTS
First	(not awarded)
Second	Ferdi Schaffers and Walter Freyn
Third	Richard Schadewell and Joachim Grasshoff
Consolation	Ferdi Schaffers and Walter Freyn
	Richard Schadewell
	6 others

The Frankfurt telephone was sometimes also referred to as the Bauhaus phone (Schadewell 1929). But what exactly was the connection between the Fuld & Company telephone, the New Frankfurt project, and the Bauhaus school? In 1927, Fuld & Company held a design competition for telephones that resulted in the Frankfurt model. Ernst May was one of the judges of this competition (Mergelsberg 2013, 9). About 250 designs were submitted, and the judges did not award a first prize (table 27.1) (H. Fuld & Co. 1928). According to a company's spokesperson in an article in *Die Form*, they did not adopt any of the models submitted to the competition, but "they gave us some ideas" (H. Fuld & Co. 1930, 220). The model submitted by Richard Schadewell (second to bottom image in fig. 27.1) was chosen as the basis for the new telephone, and technicians then adapted the shape to the parts to be housed in the apparatus.

Thus the connection between the phone, the New Frankfurt project, and the Bauhaus runs through May. He represented the city of Frankfurt on the judges' panel that selected the telephone design, which later became the standard telephone in the housing project (Neue Frankfurt n.d.). May also led the New Frankfurt project, which commissioned Walter Gropius to build a part of it, and Gropius had founded the Bauhaus school (Gropius, n.d.). Further, we believe that Harry Fuld, May, and Gropius were all progressive and would have been on the same political side in Germany at that time—and would have found it easy to work with each other.

However, we can find no connection to Marcel Breuer as stated or implied by the Kirkland Museum in Denver, the Museum of Applied Arts & Sciences in Sydney, the Museum der Dinge in Berlin, and the Bauhaus Foundation in Tel Aviv. The contemporary statement in *Die Form* by a Fuld & Company spokesperson does not mention Breuer, and the Museum of Modern Art in New York identifies only Schadewell as the designer. Sylvia Ziegner at the archives of the

Bauhaus foundation in Dessau said that they know of this so-called "Frankfurt or Bauhaus" telephone, but they do not own any material connected to it. Moreover, they do not have any material related to Schadewell in their archive. Schadewell lived in Munich, which is not near Dessau, so it appears that he was not a student at the Bauhaus school and thus not connected to Gropius or Breuer, who both were at the Bauhaus in 1927.

It is fair to conclude that Schadewell alone was responsible for the basic design and that small changes were made by Fuld & Company engineers. Judging from words of a spokesperson for Fuld & Company, the firm had an antagonistic attitude toward artists. The very first line of the statement in *Die Form* was "Wir beschäftigen keine Künstler" (We do not employ artists) (H. Fuld & Co. 1930, 220). The statement also said that another treatment of the external form by an artist was not considered, and the engineers determined the final shape. These comments would have been politically correct at that time. Regular US patents (not design patents) were awarded to Johann Schneider and Georg Heckmann for the base and handset of the Frankfurt phone, so it is likely that they made the final shape changes as well as the engineering design (Schneider and Heckmann 1930a, 1930b).

At first glance, the Frankfurt model might seem insignificant from an industrial design point of view or even off-putting—a design that is neither concave nor convex and without sex is the darndest thing! But look closely at the handset cradle. Here for the first time we see a cradle that is fully integrated with the body, rather than mounted on a perch. All future designs would incorporate integrated cradles. It is also worth noting that the model is surprisingly similar in appearance to the Bell System's No. 500 Touch-Tone telephone that would come along 35 years later (see chapter 22).

Again we can look at the handset for other design clues. Within our measurement uncertainties of about 1 millimeter and 1 degree, the important dimensions were almost exactly the same as those for the Siemens & Halske Model 26 handset, which was very similar to the handset for the Automatic Electric No. 1.

SIEMENS & HALSKE MODEL 29 HOCKENDER HUND

The Siemens & Halske Model 29 (fig. 27.2) was the first attempt in a long struggle to achieve a dial-in-handset phone for residential use (Arbenz 2020). Produced in 1929, it was designed by Otto Weeber and Otto Soldan, who were awarded US design and regular patents for the phone (Weeber and Soldan 1931, 1934). It

FIGURE 27.2 Siemens & Halske Model 29 Hockender Hund telephone, 1929. (Photo courtesy of Museumsstiftung Post und Telekommunikation.)

is noted in chapter 17 that Siemens & Halske did not employ designers at that time, but engineering organizations often had engineers in their employ who had artistic talents, and this seems to be another one of those cases.

The Model 29 became widely known as the Hockender Hund (crouching dog) because of its shape (fig. 27.3), and there were several features of this design that are noteworthy.

First, there was a switch on the bottom such that when the phone was lifted off a table, the line connection was made. When it was placed back on the table, the phone was hung up. This same format was used later by Ericsson in its more successful Ericofon (see below).

Second, in an effort to make the phone about the size of a handset, a small-diameter dial (fig. 27.4) was designed by reducing the space between the digits 1 and 0 (Rinkow 1927). To accomplish this, the finger stop 3 was movable, and its rest position is shown in the figure. Its movement was limited by pins 5 and 6, which were fastened to the base plate. Pin 10 was fastened to the finger wheel and pushed the finger stop back to its rest position when dial rotation was completed.

FIGURE 27.3 Resemblance of the Model 29 telephone to a Hockender Hund (crouching dog). (Illustration by Nicolas Waeckel.)

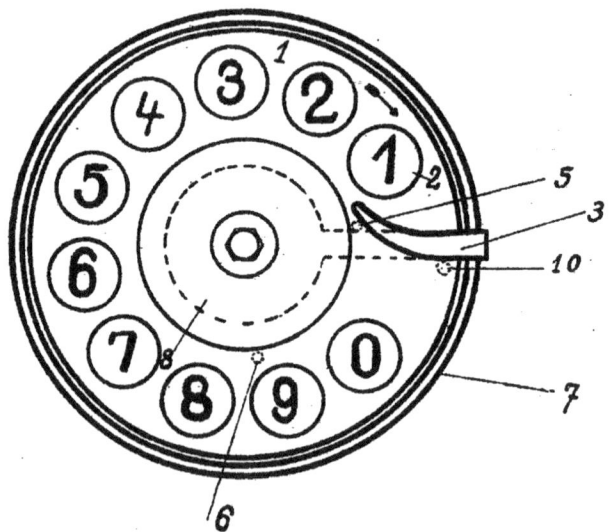

FIGURE 27.4 Patent drawing showing the movable finger stop on the Model 29 telephone. (Reprinted from Rinkow 1927.)

A similar mechanism was famously used later by Bell Labs in the design of its popular Trimline telephone (see chapter 25).

Third, it was heavy (1 lb, 7 oz; 0.65 kg) compared with the Ericofon (15 oz; 0.43 kg) and the Trimline handset (14 oz; 0.40 kg). In the effort to keep the telephone small, its base was small and the phone would easily fall over. The Ericofon, on the other hand, had a broad base and would sit up stably, but it was awkward to hold. Although a small number of the Model 29 telephones were produced, it was not introduced commercially.

SIEMENS & HALSKE MODEL 261 TROMMELWÄHLER

Hans Domizlaff joined Siemens & Halske in 1933 as a personal advisor on marketing to Carl Friedrich von Siemens, the head of the company (Arbenz 2016, 42). Domizlaff had a broad range of skills—artist, writer, painter, aviator, sailor, astronomer, brand engineer, and environmentalist—so he also became involved in product design. In switchboards of large systems, Siemens & Halske had been using a train number switch, with which numbers were dialed by pulling straight down on a numbered drum. Domizlaff adopted this concept to design a novel desk telephone, but World War II disrupted development and Domizlaff left the company.

After the war, Siemens & Halske resurrected the design, applied for a patent in 1949, and manufactured the Model 261 telephone (fig 27.5) for its private branch exchange (PBX) business (Döring 1952). The patent was issued to Fritz Döring, who probably developed details of the design, but Domizlaff's name was inexplicably omitted. The new phone became known as the Trommelwähler (drum dialer).

It was believed that the drum dialer would be easier to use repeatedly by PBX operators than a rotary dial. Apparently, this turned out not to be the case. From figure 27.5, it can be seen that the shallow and rounded recesses pose a risk of aborted and erratic dialing compared with well-defined holes in a regular rotary dial. As there is no finger stop to decisively terminate movement, it is also possible to over-pull the dial, especially if there are longer fingernails involved. This basic model was produced in both Bakelite and thermoplastic but remained in production for only five years and was never sold for residential use. Here was a case, like the outboard numbers on the Western Electric No. 500 telephone, where good design intentions did not achieve their intended purpose.

FIGURE 27.5 Siemens & Halske Model 261 Trommelwähler (drum dialer) telephone, 1950. (Photo by Arwin Schaddelee.)

During that same time period, Bell Labs also developed and patented a drum dialer in the body of the No. 302 telephone (Fassbender 2020; Madden 1953). A Western Electric prototype was built and tested within the Bell Labs facilities, but that design was abandoned also. It is noteworthy from a design point of view that both the Western Electric and the Siemens & Halske designs put the number 1 position at the bottom of the number array rather than at the top—which would have been more consistent with previous rotary dials and forthcoming Touch-Tone dials—because of the desire to use a shorter stroke for the more frequently dialed numbers.

Later in 1955, Siemens & Halske introduced a Model 282 successor to the drum dialer and it came with a standard rotary dial, yet it still had the handset lying lengthwise (fig. 27.6). In the same year, Wilhelm Pruss and Herbert Oestreich filed for a US design patent for this phone (Pruss and Oestreich 1957). After successfully completing his studies in precision engineering, Oestreich worked as an industrial designer at Siemens & Halske, as an employee at Wilhelm Wagenfeld, and freelanced when not employed by others. Later he was an industrial

FIGURE 27.6 Siemens & Halske Model 282 rotary dial telephone, 1955. (Photo by Arwin Schaddelee.)

design professor, and subsequently dean, at Kunsthochschule Kassel, the art college in Kassel, Germany. We found no further information on Pruss.

The drum dialer's successor became a resounding success. Although it was not employed by the German Post Office either, it succeeded in the extension phone and export business and by 1973, 2.7 million units had been sold. Oestreich received a gold medal for this design at the Triennale Exhibition of Modern Industrial, Architectural and Decorative Arts in Milan in 1954, and the telephone is also in the collection of the Museum of Modern Art in New York (Krause 2020, 109, 120).

SIEMENS ITALY GRILLO

The Grillo telephone (fig. 27.7) was manufactured in Italy by Siemens Italy, which is now part of Telecom Italia. Its clamshell design set Grillo apart from other telephones. Innovative features that contributed to the phone's compact size included a dial that replaced the conventional rotary finger stop mechanism with a button in each of the number holes that, when actioned, pushed a pin through the back of the dial to stop the mechanism in its correct position. The incorporation

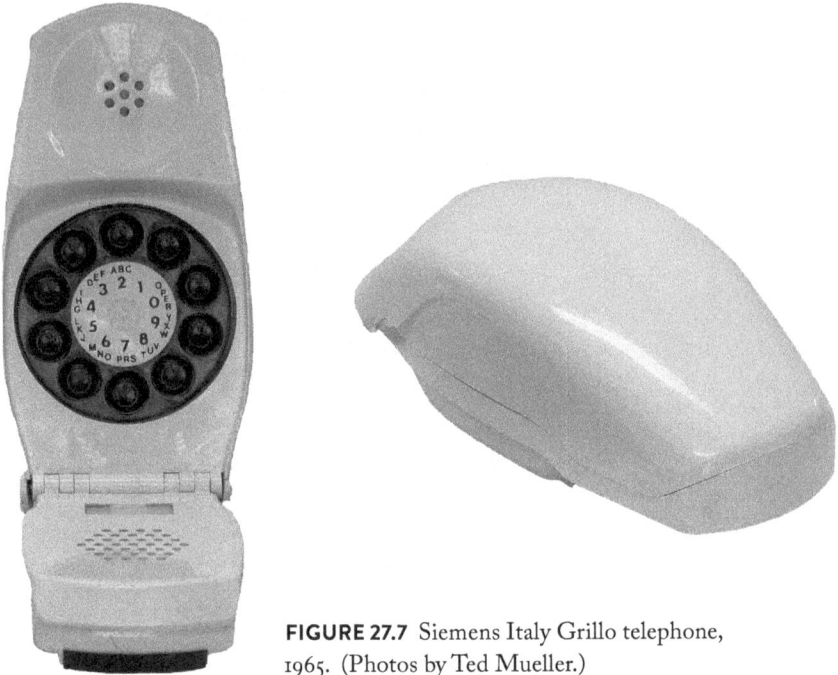

FIGURE 27.7 Siemens Italy Grillo telephone, 1965. (Photos by Ted Mueller.)

of the ringer mechanism in the wall plug rather than the phone itself and the use of a thin plastic shell also helped reduce both its size and weight. The name Grillo, which means "cricket" in Italian, was chosen because the phone's ringer sounded like a cricket.

Grillo was designed in 1965 by Richard Sapper and Marco Zanuso (Olivares 2016, 119). The design was awarded the 1967 Compasso d'Oro in Milan and the Gold Medal at the 1968 Ljubljana Biennale of Design. The Grillo is similar in form to, although larger than, the subsequent flip phone mobile telephones developed during the 1990s.

ERICSSON ERICOFON

Ericsson's project to design a one-piece telephone was begun in 1941 and progressed slowly because of the war (Blomberg 1956; Brunnström 2006, 220). For well over 40 years Ericsson had used consultants in architecture and design, but in 1943 they decided to incorporate design expertise into their organization. The new unit, called Aesthetic Design, was headed by the industrial designer Ralph Lysell. Lysell, aka Rolf Nyberg, had returned from America after nearly 16 years and had changed his surname to that of his mother's maiden name and Anglicized his first name. Based on patent awards, however, Knut Blomberg, under Lysell's direction, appears to have been the principal designer of the new one-piece telephone (Blomberg 1946; Blomberg et al. 1958; Blomberg and Lysell 1947).

The new telephone was called Ericofon (fig. 27.8), although it was also called Cobra by some because it resembled a coiled snake ready to strike. A goal of the new one-piece telephone was that it should not weigh more than the standard Bakelite handset of a typical phone. At 15 ounces (0.43 kg), the Ericofon weighed just an ounce (28 g) more than the contemporary Western Electric Type G Bakelite handset. In order to achieve this low weight, a lightweight plastic case was

FIGURE 27.8 Ericsson dial-in-handset Ericofon telephone, 1954. (Photo by Matthew Gay, courtesy of the Gregg Museum of Art & Design, North Carolina State University.)

mounted on a chassis made of an aluminum alloy and the ringer was placed in a separate wall box.

To produce a lightweight case with these unusual shapes, then recently developed thermal plastic was sprayed in a thick layer on the inside of a mold. An inside die was subsequently inserted into the base and ribs were joined to the shell by injection molding. These ribs were used to capture metal brackets on which components were fastened. An injection molded cap holding a small transmitter was glued into the opening at the top. At first the case was made of two mirrored halves that were glued together. Later, production was simplified by producing the case as a single piece using the same techniques.

Production of the Ericofon began in 1954, and the major market was thought to be hospitals, in which a patient could dial a phone without getting out of bed; the one-piece telephone seemed to be just what the doctor ordered. But the Ericofon became more widely popular and production was increased. Eventually the Ericofon was also manufactured at plants in America, Australia, and the Netherlands. The Ericofon is also in the collection of the Museum of Modern Art in New York.

CONCLUSION

T he telephone was one of the first products of the industrial revolution that was influenced by what we would call industrial design. A corded telephone was intimate, being held against one's face; it was also very visible, usually located in a living area rather than out of sight, such as a sewing machine in a sewing room or a typewriter in an office. Consequently a telephone's appearance and user friendliness were important design considerations, yet the design of these earlier telephones has been largely overlooked by historians.

Industrial design is a combination of form and function that requires designers to work with engineers. In the case of the telephone, these designers are sometimes engineers, physicians, and even physicists. Based on our investigation, the following individuals deserve recognition as important industrial designers, yet they are mostly unknown:

- William Francis Channing for the convenient handheld receiver of 1877
- Harold French Dodge for the Bell System's first handset and companion desk stand of 1930
- George Renwick Lum for the iconic No. 302 telephone of 1937
- Robert Haven Hose with Henry Dreyfuss for the No. 500 desk set of 1949 that was copied around the world

- Robert Haven Hose for the lovely Princess bedroom extension phone of 1959
- Donald Michael Genaro for the Trimline telephone of 1965 that is in the collection of the Museum of Modern Art

Henry Dreyfuss, who is of course well known, also contributed to telephone design and is now recognized as one of the pioneers of industrial design. But Dreyfuss was not a seasoned designer in 1930 when he started consulting with Bell Labs, so it would be incorrect to assume that he brought industrial design to the Bell System. Young Dreyfuss in effect served an apprenticeship in the Bell System, where he learned from their mistakes and experience. He was, in fact, very talented and eventually became so well known that he is sometimes credited for the work of others.

Perhaps it's understandable that design historians give credit to designers they know rather than digging deep to discover the true artisans. For example, Jean Heiberg, who styled the Ericsson phone of 1931, is unduly credited for giving inspiration to Dreyfuss for the No. 302 telephone. Marcel Breuer is reported to have designed the Frankfurt phone, yet he did not. And Ralph Lysell (aka Rolf Nyberg) is given credit for designing the Ericofon while Knut Blomberg, who was probably its principal designer, is slighted.

By the time of its court-ordered breakup in 1984, the Bell System had more than 100 million corded telephones in service—one for every two people in America—so the telephone was ubiquitous and its design was important. With many Nobel Prizes in physics, the Bell System is acknowledged for its technical expertise, but it should also be recognized for a formative and substantial impact on the emerging discipline of industrial design.

REFERENCES

Aldridge, A. J., E., J. Barnes, and E. Foulger. 1929. "A New C.B. Microtelephone." *The Post Office Electrical Engineers' Journal* 22 (October): 185–93.

Arbenz, Dietrich. 2009. *Vom Trommelwähler zu Optiset E: Die Geschichte der drahtgebundenen Telephone für die Wählenebenstellenanlagen von Siemens (1950-2000)*. Herbert Utz Verlag.

Arbenz, Dietrich. 2016. "Der 'Trommelwähler' ein von Hans Domizlaff konzipiertes innovatives Siemens-Telefon." *Sammler und Interessengemeinschaft für das historische Fernmeldewesen* 56 (December): 41–45.

Arbenz, Dietrich. 2020. "The Siemens 'Squatting Dog' Telephone." *Singing Wires, The Journal of Telephone Collectors International* 34 (August): 8–11.

American Society for Quality. n.d. "Harold F. Dodge." Accessed August 21, 2024. https://asq.org/about-asq/honorary-members/dodge.

AT&T. 1907. *1906 Annual Report to the Stockholders*. American Telephone & Telegraph.

AT&T. 1908. *1907 Annual Report to the Stockholders*. American Telephone & Telegraph.

AT&T. 1911. *1910 Annual Report to the Stockholders*. American Telephone & Telegraph.

AT&T. 1927. *Specifications 4810 Stations Hand Telephone Sets*. American Telephone & Telegraph.

AT&T. 1957. "Headquarters Summary." *Bell Telephone Magazine* 36 (Autumn): 56–57.

AT&T. 1967. "Adapting Products to People." *Bell Telephone Magazine* (July): 19–24.

AT&T Archives. 1992. *Events in Telecommunications History*. American Telephone & Telegraph.

Automatic Electric. 1925. "More Facts About the Monophone." *Automatic Telephone* (September–December): 66–68.

Automatic Electric. 1931. *Strowger Dial Telephones & Telephone Parts Catalogue 4014*. Automatic Electric.

Automatic Electric. 1934. *Telephones and Telephone Parts Catalogue 4055*. Automatic Electric.

Bell, Alexander G. 1876a. *Experiments Made by A. Graham Bell (Vol. I)*. Handwritten notebook. Library of Congress, Alexander Graham Bell Family Papers.

Bell, Alexander G. 1876b. Improvement in telegraphy. US Patent 174,465, filed February 14, 1876, and issued March 7, 1876.

Bell, Alexander G. 1877. Improvement in electric telegraphy. US Patent 186,787, filed January 15, 1877, and issued January 30, 1877.

Bell, Alexander G. 1878. Letter to the Capitalists of the Electric Telephone Co., March 25. Library of Congress, loc.gov, Alexander Graham Bell Family Papers.

Bell, Alexander G. 1908. Deposition of Alexander Graham Bell on April 4, 1892, to the Circuit Court of the United States, District of Massachusetts (Red Book). American Bell Telephone Co.

Bell Canada. n.d. "The Combined Hand Telephone Set in Relation to the Station Apparatus Situation of the Bell Telephone Company of Canada." Bell Telephone Co. of Canada memorandum, undated ca. 1935, Bell Archives, Montreal.

Bell Labs. 1925. "The Model Shop." *Bell Laboratories Record* 1 (September), 3–10.

Bell Labs. 1929–1932. "News of the Month." *Bell Laboratories Record* (May 1929): 377; (June 1929): 418; (March 1930): 344; (August 1930): 603; (March 1931): 347; (April 1932): iii; (May 1932): vi.

Bell Labs. 1931. "Contributors to This Issue." *Bell Laboratories Record* 10 (November): 98–100.

Bell Labs. 1943a. "Twenty-Five-Year Service Anniversaries." *Bell Laboratories Record* XXI (January): xxii–xxiv.

Bell Labs. 1943b. "Retirements." *Bell Laboratories Record* XXI (June): 383–84.

Bell Labs. 1952. "Historic Firsts: Lettered Dial." *Bell Laboratories Record* 30 (January): 12–13.

Bell Labs. 1956a. "Laboratories Introduces 5300-Type Telephone." *Bell Laboratories Record* 34 (August): 313.

Bell Labs. 1956b. "Nobel Prize Awarded to Transistor Inventors." *Bell Laboratories Record* 34 (November): 401–2.

Bell Labs. 1959. "Designers Honor 500-Type Set." *Bell Laboratories Record* 37 (April): 151.

Bello, Francis. 1954. "Fitting the Machine to Man." *Fortune Magazine* (November): 134–58.

Bennett, Arthur. F., and Charles R. Moore. 1929. Intelligence signaling apparatus. US Patent 1,719,645, filed August 20, 1925, and issued July 2, 1929.

Berliner, Emile. 1891. Combined telegraph and telephone. US Patent 463,569, filed June 4, 1877, and issued November 17, 1891.

Bernhard, Lucian. 1930. Desk stand for a hand telephone; Hand telephone. US Patents Des. 81,473, Des. 81,474, Des. 81,475, Des. 81,680, Des. 82,414, filed May 10, 1930, and issued various dates in 1930.

Billington, Henry E. 1937. Telephone desk stand. US Patent Des. 104,087, filed March 30, 1936, and issued April 13, 1937.

Black, R., and H. K. Cunningham. 1954. "Testing Telephone Usefulness." *Bell Laboratories Record* 32 (January): 25–26.

Blake, Francis. 1881. Speaking-telephone. US Patents 250,126, 250,127, 250,128, and 250,129, all issued November 29, 1881; applications for 250,126, 250,127, 250,128 filed September

15, 1881; application for 250,129 filed October 31, 1881.

Blauvelt, William G. 1922. Numbering system for automatic telephone exchanges. US Patent 1,439,723, filed September 17, 1918, and issued December 26, 1922.

Blomberg, Knut Hugo. 1946. Telephone set. US Patent 2,405,543, filed December 30, 1942, and issued August 6, 1946.

Blomberg, Knut Hugo, and Ralph Åke Gösta Lysell. 1947. Telephone handset. US Patent 2,419,388, filed October 15, 1941, and issued April 22, 1947.

Blomberg, Knut Hugo, Helge Edward Lindström, and Hans Gösta Thames. 1958. Casing for telephone instruments. US Patent 2,822,432, filed December 29, 1951, and issued February 4, 1958.

Blomberg, Hugo. 1956. "The Ericofon—the New Telephone Set." *Ericsson Review* 33: 99–109.

Blount, Nelson. 1930. Desk stand for a hand telephone. US Patents Des. 80,670, Des. 80,671, Des. 80,672, filed January 25, 1930, and issued March 1, 1930.

Blount, Nelson. 1938. Acoustic device. US Patent 2,123,177, filed September 26, 1936, and issued July 12, 1938.

Bredehoft, H. A. 1951. "New Ringer for 500-Type Telephone Set." *Bell Laboratories Record* XXIX (October): 471–74.

Brooks, John. 1975. *Telephone: The First Hundred Years*. Harper & Row.

Brown, Ralph. 1954. "Inventing and Patenting at Bell Laboratories." *Bell Laboratories Record* 32 (January): 5–10.

Brown, Robert G. 1880. Electric speaking-telephone. US Patent 224,138, filed September 29, 1879, and issued February 3, 1880.

Brown, William S. 1966. "A Decade of New Products." *Bell Telephone Magazine* 45 (Spring): 10–19.

Bruce, Robert V. 1990. *Bell: Alexander Graham Bell and the Conquest of Solitude*. Cornell University Press.

Brunnström, Lasse. 2006. *Telefonen en Designhistoria*. Bokförlaget Atlantis.

Burlin, James N., and Robert H. Hose. 1958. Telephone stand. US Patent Des. 182,498, filed December 17, 1956, and issued April 15, 1958.

Campbell, George Ashley. 1937. *The Collected Papers of George Ashley Campbell*. American Telephone & Telegraph.

Channing, William F. 1877. Letter to A. Graham Bell, July 16. Library of Congress, loc. gov, Alexander Graham Bell Family Papers.

Channing, William F., and Moses G. Farmer. 1857. Electromagnetic fire alarm telegraph for cities. US Patent 17,355, issued May 19, 1857.

Clarke, Harry R. 1926. Telephone hand set. US Patent 1,601,060, filed February 7, 1923, and issued September 28, 1926.

Clarke, René. 1930. Desk stand for a hand telephone; Hand telephone. US Patents Des. 81,476, Des. 81,477, filed May 10, 1930, and issued July 1, 1930.

Clarke, Sally. 1998. "Negotiating Between the Firm and the Consumer: Bell Labs and the Development of the Modern Telephone." *The Modern Worlds of Business and Industry*, 161–82, edited by Karen R. Merrill. Brepols.

Cobb, L. J. 1952. "Handset for the 500-Type Set." *Bell Laboratories Record* 30 (August): 317–20.

Colorado Springs Gazette Telegraph. 1982. "Obituary: George R. Lum," January 3, p. 4A.

Colpitts, Edwin H. 1928. Letter to H. P. Charlesworth at Bell Telephone Laboratories, December 27. AT&T Archives and History Center, Warren, NJ.

Colpitts, Edwin H. 1929a. Letter to H. P. Charlesworth at Bell Telephone Laboratories, March 19. AT&T Archives and History Center, Warren, NJ.

Colpitts, Edwin H. 1929b. Memorandum to H. P. Charlesworth at Bell Telephone Laboratories, July 3. AT&T Archives and History Center, Warren, NJ.

Connolly, M. Daniel, Thomas A. Connolly, and Thomas J. McTighe. 1879. Improvement in automatic telephone-exchanges. US Patent 222,458, filed September 10, 1879, and issued December 9, 1879.

Curtis, Alfred S. 1925. Telephone set. US Patent 1,560,768, filed December 24, 1920, and issued November 10, 1925.

Daugherty, Greg. 2023. "The Rise and Fall of Telephone Operators." The History Channel. Updated October 2. https://www.history.com/news/rise-fall-telephone-switchboard-operators.

de Forest, Lee. 1908. Space telegraphy. US Patent 879,532, filed January 29, 1907, and issued February 18, 1908.

Deininger, R. L. 1960. "Human Factors Engineering Studies of the Design and Use of Pushbutton Telephone Sets." *Bell System Technical Journal* 39 (July): 995–1012.

Dilts, Marion M. 1931. "The Golden Section." *Bell Laboratories Record* 10 (November): 97–98.

Dodge, Harold F. 1928. Transmitter mouthpiece. US Patent 1664852, filed December 30, 1922, and issued April 3, 1928.

Dodge, Harold F. 1931. "Squares and Rectangles." *Bell Laboratories Record* 10 (November): 93–96.

Döring, Fritz. 1952. Fernsprechstation mit Nummernschalter. DE Patent 852,400, filed July 8, 1949, and issued October 13, 1952.

Dreyfuss, Henry. 1935. Air conditioning control. US Patent Des. 96,822, filed July 19, 1935, and issued September 10, 1935.

Dreyfuss, Henry. 1938. Combined tractor radiator cover and hood. US Patent Des. 112,365, filed September 2, 1938, and issued November 29, 1938.

Dreyfuss, Henry. 1939. Locomotive. US Patent Des. 116,180, filed September 28, 1938, and issued August 15, 1939.

Dreyfuss, Henry. 1948. Hand telephone. US Patent Des. 151,614, filed June 24, 1947, and issued November 2, 1948.

Dreyfuss, Henry. 1955. *Designing for People.* Simon and Schuster.

Dreyfuss, Henry. 1963. Pushbutton telephone desk stand. US Patent Des. 197,067, filed October 17, 1962, and issued December 10, 1963.

Dreyfuss, Henry. 1967. *Designing for People.* Grossman Publishers.

Dreyfuss, Henry. 1969. *A Record of Personal and Business Vital Statistics.* Unpublished "Brown Book" in the Smithsonian Libraries collection at Cooper Hewitt Library.

Dreyfuss, Henry. 1974. *Designing for People.* Viking Press.

Dreyfuss, Henry. 2003. *Designing for People.* Allworth Press.

Dreyfuss, Henry, and Robert H. Hose. 1949. Desk stand for a hand telephone. US Patent Des. 153,927, filed February 11, 1948, and issued May 31, 1949.

Dreyfuss, Henry, Robert H. Hose, and James N. Burlin. 1959. Telephone set. US Patent Des. 184,307, filed June 24, 1958, and issued January 27, 1959.

Dreyfuss, Henry, Lionel W. Mosing, and Robert E. Prescott. 1965a. Combined telephone handset and stand. US Patent Des. 202,787, filed September 21, 1964, and issued November 9, 1965.

Dreyfuss, Henry, Lionel W. Mosing, and Robert E. Prescott. 1965b. Telephone handset. US Patent Des. 202,788, filed September 21, 1964, and issued November 9, 1965.

Dubas, Rita. 2006. *Shirley Temple: A Pictorial History of the World's Greatest Child Star.* Applause Books.

Eaton, George R. 1935. Microphone. US Patent 2,014,427, filed April 11, 1933, and issued September 17, 1935.

Eaton, George R., and Royal Horn. 1931. Hand telephone. US Patent Des. 83,599, filed January 14, 1931, and issued March 10, 1931.

Erickson, Clifford E. 1948. Telephone desk unit. US Patent Des. 151,121, filed April 19, 1946, and issued September 28, 1948.

Estreich, Bob, and Jan Verhelst. n.d. "The Bell Telephone Manufacturing Company of Antwerp, Belgium." Bob's Old Phones, website of the Australasian Telephone Collectors Society Inc. Accessed August 22, 2024. http://www.telephonecollecting.org/Bobs%20phones/Pages/Bell%20Antwerp/Bell%20Antwerp.html#:~:text=On%20 26th%20April%201882%20the,returned%20to%20the%20United%20States.

Fagen, M. D., ed. 1975. *A History of Engineering and Science in the Bell System: The Early Years (1875–1925).* Bell Telephone Laboratories.

Fassbender, Paul. 2011. "Western Electric D-95647 Sold for $3,150." *Singing Wires: The Journal of Telephone Collectors International* 25 (December): 1, 8.

Fassbender, Paul. 2020. "Why Build a Linear Rotary Dial?" *Singing Wires: The Journal of*

Telephone Collectors International 34 (March, April, and May): 1, 6–7.

Ferebee, Ann. 1970. *A History of Design from the Victorian Era to the Present.* Van Nostrand Reinhold.

Flinchum, Russell A. 1997. *Henry Dreyfuss, Industrial Designer: The Man in the Brown Suit.* Rizzoli International Publications.

Flinchum, Russell A., and Ralph O. Meyer. 2017. "Henry Dreyfuss and Bell Telephones." *Winterthur Portfolio* 51 (Winter): 173–200.

Flinchum, Russell A., and Ralph O. Meyer. 2022. *Henry Dreyfuss: Designing for People.* State University of New York Press.

Freshwater, Robert. 2022. "Evolution of British Post Office Telephones." The Telephone File. Last revised October 30. https://www.britishtelephones.com/histbpoteles.htm.

Genaro, Donald Michael, and John Niel McGarvey. 1974. Telephone handset. US Patent Des. 229,837, filed April 13, 1973, and issued January 8, 1974.

Gertner, Jon. 2012. *The Idea Factory: The Bell Labs and the Great Age of American Innovation.* Penguin Press.

Gherardi, Bancroft, and F. B. Jewett. 1930. "Telephone Communication System of the United States." *Bell System Technical Journal* 9 (January): 1–100.

Gray, Christopher. 2000. "Streetscapes/AT&T Headquarters at 195 Broadway; A Bellwether Building Where History Was Made." *New York Times*, April 23. https://www.nytimes.com/2000/04/23/realestate/streetscapes-t-headquarters-195-broadway-bellwether-building-where-history-was.html.

Grönwall, G. 1933. "The New Ericsson Telephone." *Ericsson Review* 10: 4–11.

Gropius, Walter. n.d. "Settlement Lindenbaum." *Architectuul.* Accessed August 22, 2024. https://architectuul.com/architecture/settlement-lindenbaum.

Grüger, W., and H. Van Holst. 1963. "Assistant Type Telephone Set." *Electrical Communication* 38: 230–39.

Hadlaw, Jan. 2017. "The Modern American Telephone as a Contested Technological Thing, 1920–39." In *Encountering Things: Design and Theories of Things*, 133–48, edited by Leslie Atzmon and Prasad Boradkar. Bloomsbury.

Hadlaw, Janin. 2004. "Communicating Modernity: Design, Representation, and the Making of the Telephone." PhD thesis, Simon Fraser University.

Hall, Norris R. 1975. "Design Line Decorative Telephones: Just the Beginning." *Bell Laboratories Record* 53 (May): 235–41.

Hanson, B. L. 1983. "A Brief History of Applied Behavioral Science at Bell Laboratories." *Bell System Technical Journal* 62 (July–August): 1571–90.

Harris, Gale. 2006. *American Telephone & Telegraph Company Building*, Report LP- 2194, July 25. City of New York, Landmarks Preservation Commission. https://s-media.nyc.

gov/agencies/lpc/lp/2194.pdf.

Hart, Rita. 1930. Letter to N. Blount at Bell Telephone Laboratories, August 27. AT&T Archives and History Center, Warren, NJ.

Hayes, Hammond V. 1892. Telephone circuit. US Patent 474,323, filed January 13, 1892, and issued May 3, 1892.

H. Fuld & Co. 1928. "Wettbewerb um Entwürfe für Fernsprechapparate." *Zentralblatt der Bauverwaltung* 48 (January): 29–30.

H. Fuld & Co. 1930. "Die Angaben der Firma H. Fuld & Co., Frankfurt a.M. lauten." *Die Form, Zeitschrift für Gestaltende Arbeit* (April 15): 220–21.

H. Fuld & Co. 1932. *Fernsprecher: Sammel No. 7011.* Catalog of H. Fuld & Co.

Hiesinger, Kathryn B., and George H. Marcus. 1993. *Landmarks of Twentieth-Century Design: An Illustrated Handbook.* Abbeville.

Hubbard, Gardiner G. 1877. "Declaration of Trust," July 28. US Department of the Interior.

Hunnings, Henry. 1881. Telephone-transmitter. US Patent 250,250, filed September 30, 1881, and issued November 29, 1881.

Huxham, T. S. 1939. "Molded Telephone Apparatus Design." *Bell Laboratories Record* 17 (March): 212–16.

International Telecommunications Union. 1988. "Technical Features of Push-Button Telephone Sets, ITU-T Recommendation Q.23." Fascicle VI.1 of the *Blue Book.* International Telecommunications Union.

Jenkins, R. T. 1957. "Distinctive Ringing Signals for the 500 Set." *Bell Laboratories Record* 35 (September): 348–52.

Jensen, Gustav B. 1930. Hand telephone. US Patent Des. 81,539, filed May 10, 1930, and issued July 8, 1930.

Jewett, F. B. 1935. Letter to Edgar S. Bloom at Western Electric Co., May 28. AT&T Archives and History Center, Warren, NJ.

Johnson, Rossiter, and John Howard Brown, eds. 1904. "Channing, William Francis." In *The Twentieth Century Biographical Dictionary of Notable Americans.* Vol. 2. Boston Biographical Society.

Jones, R. L. 1930. Memorandum to H. P. Charlesworth at Bell Telephone Laboratories, February 18. AT&T Archives and History Center, Warren, NJ.

Jones, W. C. 1938. "Instruments for the New Telephone Sets." *Bell System Technical Journal* 17 (July): 338–57.

Jones, W. C., and A. H. Inglis. 1932. "The Development of a Handset for Telephone Stations." *The Bell System Technical Journal* 11 (April): 245–63.

Kåell, Å. 1964. "Dialog—the New Telephone." *Ericsson Review* 41: 136–46.

Kelly, Howard A., and Walter L. Burrage. 1920. "Channing, William Francis." In *American*

Medical Biographies. Norman, Remington.

Kiesel, Walter C. 1922. Telephone hand set. US Patent 1,425,977, filed December 8, 1919, and issued August 15, 1922.

King, Douglas H., and Harold T. Martin. 1930. Wall mounting for a hand telephone. US Patent Des. 82,505, filed June 13, 1930, and issued November 11, 1930.

King, Douglas H. 1932. *Report on Artists' Models—Bell Labs Case 34648*, August 9. AT&T Archives and History Center, Warren, NJ.

Krause, Christoph T. M. 2020. *The Drum Dial: The Forgotten Telephone Luminary of the Federal German Foundation Years.* Tredition.

Krumreich, Charles L., and Lionel W. Mosing. 1966. "The Evolution of a Telephone." *Bell Laboratories Record* 44 (January): 9–14.

Labaugh, James M. 1930. Casing for a telephone substation set. US Patents Des. 82,387 and Des. 82,613, filed June 4, 1930, and issued on different dates in 1930.

Legg, V. E. 1939. "Survey of Magnetic Materials and Applications in the Telephone System." *Bell System Technical Journal* 18 (July): 438–64.

Lindquist, Emory. 1957. "The Invention and Development of the Dial Telephone." *Kansas Historical Quarterly* 23 (Spring): 1–8.

L.M. Ericsson Telephone Co. 1931. "The New Ericsson Subscriber's Automatic Telephone 1931 Model." *Ericsson Review* 8: 266.

Lum, George R. 1932. Desk stand for a hand telephone. US Patent Des. 88,804, filed October 20, 1932, and issued December 27, 1932.

Lum, George R. 1934. Telephone substation apparatus. US Patent 1,965,423, filed May 19, 1932, and issued July 3, 1934.

Lum, George R. 1935a. Desk stand for a hand telephone. US Patent Des. 95765, filed March 27, 1935, and issued May 28, 1935.

Lum, George R. 1935b. Hand telephone. US Patent Des. 95,915, filed April 25, 1935, and issued June 11, 1935.

Lum, George R. 1935c. Telephone substation apparatus. US Patent 2,008,287, filed October 20, 1932, and issued July 16, 1935.

Lum, George R. 1938. Support for base plates. US Patent 2,137,021, filed March 27, 1935, and issued November 15, 1938.

Lum, George R. 1949. Wall mounting for a hand telephone. US Patent Des. 152,276, filed December 9, 1947, and issued January 4, 1949.

Lupton, Ellen. 2014. *Beautiful Users: Designing for People.* Princeton Architectural Press.

Lurk, Hans-J. 1934. "Ermittlung der Masse eines der häufigsten Kopfform angepassten Handapparates unter Zuhilfenahme einer neuartigen Messapparatur." *Zeitschrift für*

Fernmeldetechnik, Werk-und Gerätebau 4 (April 18): 49–52.

Madden, James J. 1953. Linear dial call transmitter. US Patent 2,635,230, filed November 13, 1950, and issued April 14, 1953.

Manson, Ray H. 1934. "Combining Beauty with Utility in Hand Set Telephones." *Hand Set Telephone Bulletin No. 1*, draft, September 10. Strombreg-Carlson Telephone Manufacturing Co.

McFadden, Paul, ed. 1994. Untitled article. *Singing Wires: The Journal of Telephone Collectors International* 8 (March): 8.

Mergelsberg, Günther. 2013. *Die Telefone des Herrn Fuld.* Sammler- und Interessen-Gemeinschaft für das historische Fernmeldewesen e.V.

Meyer, Ralph O. 2005. *Old-Time Telephones: History, Design, Technology, Restoration.* 2nd ed. Schiffer.

Meyer, Ralph O. 2018. *Old-Time Telephones: History, Design, Technology, Restoration.* 3rd ed. North Carolina State University Libraries. https://repository.lib.ncsu.edu/handle/1840.20/35318.

Meyer, Ralph O. 2020. "Bell, Watson, Soft Iron, and the Insight That Commercialized the Magneto Telephone." *Proceedings of the IEEE* 108 (December): 2311–20.

Meyer, Ralph O. 2022. *Naked Physics of the Telegraph, Telephone, and Radio.* Kindle Direct.

Miller, Kempster B. 1933. *Telephone Theory and Practice: Manual Switching and Substation Equipment.* McGraw-Hill.

Moore, Charles R. 1925. Telephone transmitter. US Patent 1,565,581, filed October 3, 1921, and issued December 15, 1925.

Morey, C. R., and T. C. Oehne, Jr. 1908. *Lightning Arresters and Schemes for Testing.* Armour Institute of Technology.

Mosing, Lionel W. 1961. Combined telephone handset and stand. US Patent Des. 189,877, filed May 11, 1960, and issued March 14, 1961.

Museumsstiftung Post und Telekommunikation. 2001. *Telefone 1863–2000,* Museumsstiftung Post und Telekommunikation.

Neue Frankfurt. n.d. *Das Frankfurter Register* 6. Undated ca. 1930.

Obergfell, Herbert F. 1925. Combined ringer box and telephone desk stand. US Patent Des. 68,929, filed October 10, 1925, and issued December 1, 1925.

Obergfell, Herbert F. 1932. Telephone desk stand. US Patent Des. 86,263, filed May 18, 1931, and issued February 16, 1932.

Obergfell, Herbert F. 1934a. Telephone handset. US Patent 1,971,499, filed May 2, 1932, and issued August 28, 1934.

Obergfell, Herbert F. 1934b. Wall telephone. US Patent Des. 94,158, filed November 9, 1934, and issued December 25, 1934.

Olivares, Jonathan, ed. 2016. *Richard Sapper.* Phaidon Press.

Paine, Albert Bigelow. 1921. *In One Man's Life: Being Chapters from the Personal & Business Career of Theodore N. Vail.* Harper & Brothers.

Pleasance, Charles A. 1989. *The Spirit of Independent Telephony.* Independent Telephone Books.

Pocock, L. C., and L. Schreiber. 1936. "A Molded Bakelite Set with a New Microtelephone." *Electrical Communication* 14 (April): 261–69.

Polhemus, Fred. 2015. *Arthur Shilstone: A Lifetime of Drawing & Painting.* Tide-mark Press.

Prescott, George B. 1884. *Bell's Electric Speaking Telephone: Its Invention, Construction, Application, Modification and History.* D. Appleton & Co.

Pruss, Wilhelm, and Herbert Oestreich. 1957. Combined telephone desk stand and handset. US Patent Des. 179,778, filed August 29, 1955, and issued February 26, 1957.

Purcell, William F. H. 1982. "Notes on Industrial Design." Copy given to author by Purcell.

Rinkow, Otto. 1927. Nummershalter für selbsttätig betriebene Fernsprechanlagen. DE Patent 439,970, filed May 5, 1925, and issued January 25, 1927.

Robertson, J. H. 1948. *The Story of the Telephone: A History of the Telecommunications Industry in Britain.* Scientific Book Club.

Roosevelt, Hilborne L. 1879. Improvement in telephone-switches. US Patent 215,837, filed October 3, 1877, and issued May 27, 1879.

Schadewell, Richard. 1929. *Frankfurt ('Bauhaus') Telephone.* Object No. 131.2010. Museum of Modern Art, New York. https://www.moma.org/collection/works/134285.

Scharringhausen, William H. 1931. Telephone desk set. US Patent 1,788,747, filed November 22, 1927, and issued January 13, 1931.

Schenker, L. 1960. "Push Button Calling with a Two-Group Voice-Frequency Code." *Bell System Technical Journal* 39 (January): 235–55.

Schneider, Johann, and Georg Heckmann. 1930a. Microtelephone. US Patent 1,760,566, filed February 2, 1929, and issued May 27, 1930.

Schneider, Johann, and Georg Heckmann, 1930b. Telephone instrument. US Patent 1,762,641, filed July 10, 1929, and issued June 10, 1930.

Schreiber, L. 1935. "Developments in Subscriber Sets." *Electrical Communication* 14 (July): 21–34.

Scribner, Charles E. 1901. Telephone-circuit. US Patent 669,710, filed November 13, 1897, renewed September 4, 1900, and issued March 12, 1901.

Sentenne, Charles J. 1965. "Introducing: The Trimline Telephone." *Bell Telephone Magazine* 44 (Autumn): 8–11.

Siemens Brothers & Co. 1931. "The Manufacture of the Neophone." *Engineering Supplement to the Siemens Magazine* 77 (October): 1–6.

Spencer, H. J. C., and F. A. Wilson. 1959. "The New 700-Type Table Telephone—Telephone No. 706." *The Post Office Electrical Engineers' Journal* 52 (April): 1–12.

Stromberg-Carlson. 1940. Telephones, Switchboards, Telephone Parts and Accessories (a loose-leaf catalog). Stromberg-Carlson.

Taggart, Charles Johnson, and Elizabeth J. Van Allen. 2021. "Western Electric." Encyclopedia of Indianapolis. https://indyencyclopedia.org/western-electric/.

Telephony Publishing Corp. 1935. "Automatic Electric Co.'s New Telephone Transmitter." *Telephony* 109 (October 5): 50–51.

Telephony Publishing Corp. 1938. "Shirley Temple Gets 100,000th Telephone." *Telephony* 115 (September 3): 14.

Telephony Publishing Corp. 1945. "A New Type of Telephone Comes to Main Street." *Telephony* 128 (March 3): 5

Telephony Publishing Corp. 1951a. "Western Electric and ITT Sign Licensing Agreement." *Telephony* 140 (June 23): 42.

Telephony Publishing Corp. 1951b. "IT&T Acquires Interest in Kellogg Company." *Telephony* 141 (August 18): 42.

Thompson, George K. 1924a. Desk stand for hand telephones. US Patent Des. 65,204, filed December 28, 1922, and issued July 15, 1924.

Thompson, George K. 1924b. Telephone desk set. US Patent 1,508,424, filed December 28, 1922, and issued September 16, 1924.

Thompson, George K. 1925. Hand Telephone. US Patent Des. 66,991, filed December 31, 1923, and issued April 7, 1925.

Thompson, George K., and Alfred U. Harper. 1922. Hand telephone. US Patent Des. 60,180, filed October 15, 1920, and issued January 3, 1922.

Tilley, Alvin. 1953. Untitled notes, May. Henry Dreyfuss Archive, Cooper Hewitt, Smithsonian Design Museum, New York.

Vassos, John. 1930. Desk stand for a hand telephone; Hand telephone. US Patents Des. 81,510, Des. 81,511, Des. 81,512, Des. 81,513, Des. 81,562, Des. 82,068, filed May 10, 1930, and issued various dates in 1930.

Wallance, Don. 1956. *Shaping America's Products*. Reinhold.

Watson, Thomas A. 1878. Telephone call-signal apparatus. US Patent 202,495, filed October 11, 1877, and issued April 16, 1878.

Watson, Thomas A. 1926. *Exploring Life: The Autobiography of Thomas A. Watson*. D. Appleton & Co.

Watson, Thomas A. 1940. *The Birth and Babyhood of the Telephone*. Information Department, American Telephone & Telegraph.

Weeber, Otto. 1928. Hand telephone. US Patent 1,688,477, filed September 11, 1925, and issued October 23, 1928.

Weeber, Otto, and Otto Soldan. 1931. Desk telephone. US Patent Des. 85,679, filed October 12, 1931, and issued December 1, 1931.

Weeber, Otto, and Otto Soldan. 1934. Telephone set. US Patent 1,944,701, filed July 8, 1932, and issued January 23, 1934.

Western Electric. 1929. *Telephone Apparatus and Supplies No. 7*. Western Electric.

Western Electric. 1935a. *From the Far Corners of the Earth*. Western Electric.

Western Electric. 1935b. *Telephone Apparatus and Cable Catalog No. 9*. Western Electric.

Western Electric. 1939. *Telephone Apparatus and Cable Catalog No. 10*. Western Electric.

Western Electric. 1978. *Hawthorne—Its Life and People*. Western Electric.

White, Anthony C. 1892. Telephone. US Patent 485,311, filed March 24, 1892, and issued November 1, 1892.

Wilson, Linton. 1937. "Gustav Jensen." *Pencil Points* (March): 133–50.

INDEX

ABOUT THE AUTHORS

Ralph O. Meyer is a physicist with publications in the fields of tele-communications and nuclear power safety. He has a PhD from the University of North Carolina at Chapel Hill and a Phi Beta Kappa key from the University of Kentucky. He is the author of *Old-Time Telephones* (1994, 2005, and 2018), *Naked Physics of the Telegraph, Telephone, and Radio* (2022), and coauthor of *Henry Dreyfuss: Designing for People* (2022).

Russell A. Flinchum is an associate professor of industrial design at North Carolina State University. He has a PhD from the Graduate Center of the City University of New York and a Phi Beta Kappa key from the University of North Carolina at Chapel Hill. He is the author of *Henry Dreyfuss, Industrial Designer: The Man in the Brown Suit* (1997) and *American Design* (2008), and coauthor of *Henry Dreyfuss: Designing for People* (2022).